KU-083-222

ALLE · ZEIT · WACH
1842

Alexey V. Zhirmunsky
Victor I. Kuzmin

Critical Levels in the Development of Natural Systems

With 89 Figures and 33 Tables

Springer-Verlag Berlin Heidelberg NewYork
London Paris Tokyo

Acad. Prof. Dr. Alexey V. Zhirmunsky

Institute of Marine Biology
Far East Branch
Academy of Sciences
SU-Vladivostok 690022

Prof. Dr. Victor I. Kuzmin

Moscow Institute of Radioengineering
Electronics and Automation
SU-Moscow 117454

ISBN 3-540-18409-0 Springer-Verlag Berlin Heidelberg New York
ISBN 0-387-18409-0 Springer-Verlag New York Berlin Heidelberg

Library of Congress Cataloging-in-Publication Data.
Zhirmunskiĭ, Alekseĭ Viktorovich.
Critical levels in the development of natural systems.
Bibliography: p. Includes index. 1. Biological systems. 2. Biological systems – Mathematical models. I. Kuz'min, Viktor Ivanovich. II. Title.
QH313.Z4513 1988 574'.01 87-35569
ISBN 0-387-18409-0 (U.S.)

Printed in Germany

Typesetting: K+V Fotosatz GmbH, Beerfelden
Offsetprinting: Kutschbach, Berlin. Bookbinding: Lüderitz & Bauer, Berlin
2131/3020-543210

Preface

This book is a revised and considerably enhanced version of the Russian publication *Critical Levels in the Developmental Processes of Biological Systems* (Zhirmunsky and Kuzmin 1982). Herein we have shown certain quantitative patterns in the forming of a sequence of critical points in the development of biological systems. It was found that for processes of the allometric type (the dynamics of these is described by power function) the correlation between the critical values of arguments at two consecutive critical points does not exceed the constant equal to e^e. On this basis, the law of critical levels of the allometric development of systems was formulated: *in developing systems there occur among critical levels such as have a correlation of consecutive values equal to* e^e (Kuzmin and Zhirmunsky 1980b).

An analysis of extensive experimental material concerned with the development of biological systems from cell to biocenosis has shown that this critical correlation (e^e) characterizes ranges between critical levels in the processes of development of these systems. Besides, it has been demonstrated that in processes of development smaller critical correlations (e-fold) are also substantially presented, for instance, in processes of the exponential type. A study of synchronization conditions of evenly spaced critical boundaries and critical boundaries of the exponential type (the limits of which have an e-fold correlation) has resulted in a notion of an elementary *link* or a cell of *development* in which an allometric stage (characterized by the e^e correlation) as well as a reconstruction phase, which is followed by a qualitatively new cycle of development, are distinguished.

An analysis of critical boundaries in the development of the Earth's crust, carried out jointly by the well-known geologists, academician Boris S. Sokolov and Vasily D. Nalivkin, corresponding-member of the USSR Academy of Sciences, have shown that boundaries of the geological history correspond to those of the cell of development.

It turned out that for constructing a complete system of boundaries of the geological time scale, with evaluation of their levels of significance, it would be enough to know two initial values: the duration of the galactic year and the age of one of the most strongly manifested boundaries.

The use of the revealed critical constants enabled us to describe qualitatively similar ranges of changing characteristics of different developing systems.

Research carried out after the publication of the book in 1982 made it possible to introduce a hierarchy of developmental processes and bring the hierarchy of critical constants in line with it. Consideration of extensive experimental material has shown a unity of the rhythmical structure of the solar system with biological systems from the aspects both of time and of space. Datum points of synchronization of cosmic rhythms, of bodies of the solar system, and of biological systems, are determined by a set of critical constants we have established (Kuzmin and Zhirmunsky 1980a, b; Zhirmunsky and Kuzmin 1986).

The results mentioned above have required a reconstruction and a new "organization" of our book, and its revised edition with a title reflecting more completely the extended class of phenomena studied, is presented herewith for the verdict of our readers.

Moscow, Spring 1988

A. Zhirmunsky
V. Kuzmin

Foreword to the Book "Critical Levels in Processes of Development of Biological Systems"

This book by the biologist A. V. Zhirmunsky and the specialist in system modeling V. I. Kuzmin is devoted to the developmental processes in biological systems on different levels of organization of living matter. The problem considered is a central one both in biology and in natural sciences in general. The development of natural systems is characterized, on the one hand, by a great variety of specific features and the mechanisms of their realization and, on the other, by certain general tendencies which were discovered long ago in philosophy and the natural sciences. The book is devoted to a quantitative analysis of the general laws (or regularities) of developmental processes.

For a long time in natural sciences there have been antagonistic trends: one giving preference to evolutionary tendencies in developmental processes, the other concentrating attention on periodical revolutionary changes ("catastrophes"). Lately, a consolidation of these points of view is being brought about in which evolution and revolution are taken as successive developmental phases connected with a transition from quantitative changes to qualitative ones.

Until quite recently, the studies of developmental processes were of a descriptive nature or were based on evolutionary models. But certain attempts were made – but not completed – to use uninterrupted models at separate stages of the system's development. Ranges of applicability for the models were determined through analysis of the factual morphofunctional material. The questions of how specific are the starting points of a new phase of system development, and whether some general regularities exist in the changing of developmental phases, are still open to argument. This is the very problem the book is devoted to.

The peculiarity of the author's approach to the problem is an account of relaxation reactions to the formation of current characteristics of the developmental process. The book presents factors and regularities of the formation of developmental stages and, further, a mathematical model is introduced whose results are examined on ample material.

The main result presented by the authors for discussion is a determination of some constant correlations between characteristics of developing systems at a moment of their changing the developmental character. If the authors are right, the conclusion will be of extremely great, both theoretical and practical, importance. For instance, a knowledge of quantitative values of critical abundances might be a theoretical basis for taking a number of measures connected with environmental control and the management of biological resources. The regularities established could be of considerable interest also for geologists, particularly in making a unified geochronological scale.

Together with further testing of the results obtained in biology, an analysis of analogous phenomena beyond the borders of biology and geology is thought to be of significant interest.

Academician M. S. Gilyarov
Academician B. S. Sokolov

Contents

Introduction

The development of organisms, their communities, the biosphere, the Earth and the solar systems has always attracted the attention of naturalists. What are the regularities in the development of these and other natural systems in time and space? What do the functioning and development of systems on different levels of organization and integration have in common, and what are the principal differences between them? How do slow (evolutionary) and fast (revolutionary, spasmodic) developmental stages relate? These problems are of great theoretical interest and answers to them may also be of practical importance.

When analyzing the growth data of marine molluscs and other animals, the authors paid attention to the changes in the regime of animal development and began to study their relations. In the process of studying, a regular change from allometric growth periods to shorter periods of reorganization was observed. By the use of the theory of relaxation processes we were able to derive mathematical equations and offer a model of change from evolutionary allometric developmental stages to critical periods.

To simulate growth, two main approaches are used. The first and more usual approach consists in attempts to introduce a unified growth curve that can describe the whole period of development (Von Bertalanffy 1968; Vinberg 1968; and others).

The second is to single out the stages of the development where they occur within the framework of constant (or close to constant) exogenous and endogenic conditions (Brody 1927; Schmalhausen 1935).

Ostroumov (1912, 1918) noted that growth was heterogenous during the whole life of the organism and had to be divided into separate phases, at each of which various factors played a determinative part. In detail, the possibilities of this direction were investigated by Schmalhausen (1935) within the framework of the parabolic growth model. In this case, the growth process is described by a power equation:

$$y = aT^{B} ,$$

where y is the size of the system, T is its age, and a and B are constants. The value of B-constant shows by how many percents the function of y changes when the argument changes by 1%.

When growth-limiting factors change, a spasmodic change in the value of B-constant occurs, as Schmalhausen (1935) showed. An example of such a change in the B-constant value is a process of growth of the human body mass considered by Schmalhausen (1935, Fig. 1). One can see that values of relative increase of

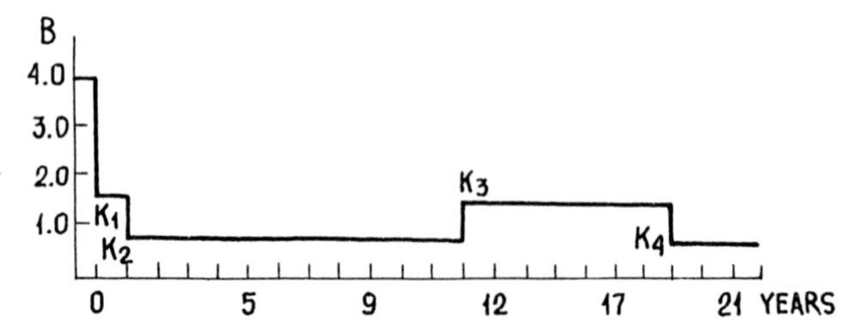

Fig. 1. Changes in intensity of boys' growth with age. *Abscissa* age after birth; *Ordinate* the constant of growth – an index of allometric dependence extent *B*, characterizing the percentage of mass increase per 1% of age increase; $K_1 - K_4$ critical ages at which the growth constant changes (According to Schmalhausen 1935)

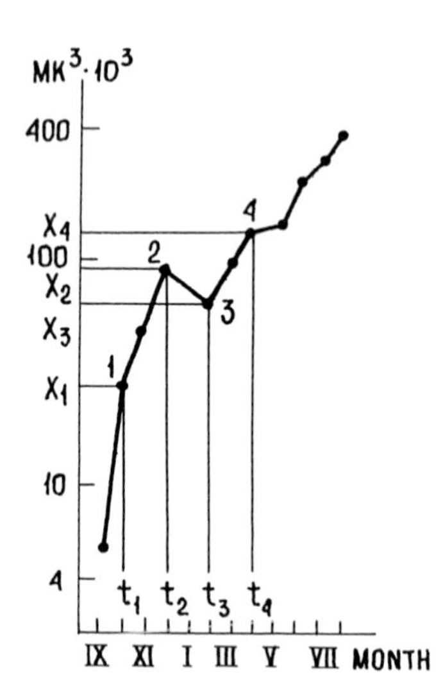

Fig. 2. Change in volume of the sea urchin *Strongylocentrotus nudus* eggs with age. *Abscissa* age (months of annual cycle); *Ordinate* egg volumes (the scale is logarithmic); *1–4* critical points; $t_1 - t_4$ critical ages; $x_1 - x_4$ – critical levels. Development between points *2* and *3* is the critical period (reconstruction phase) (According to Gnezdilova et al. 1976)

mass per unit of relative increase of age (B) are constant at certain stages, and at the moment of birth and at the ages of 1, 11, and 19 years they change by leaps.

At these same ages, in the human organism, changes resulting in radical reorganizations of metabolic processes occur: transition from the embryonic period, when feeding comes from the mother organism, to the postembryonic lactation period; then, a transition to independent feeding; a stimulation of growth dependent on sexual maturation; and cessation of growth.

Critical states in the development of the system (reorganizations) may be demonstrated on the curve of the dependence of the sea urchin *Strongylocentrotus nudus* ova volumes on the age of cells (Fig. 2). Points of break in the growth curve (points 1–4) are seen; we shall call them critical points. Corresponding ages ($t_1 - t_4$) will be called critical ages or critical arguments, and corresponding values of functions ($x_1 - x_4$) critical levels. The transitional stage between two successive growth periods of ova sizes, at which these sizes are diminishing, will be called the critical period or the phase of reorganization. As one can see in Fig. 2, two successive phases of ova growth 1–2 and 3–4 are interrupted by the phase of reorganization 2–3.

Thus, a possibility to single out by morphofunctional features phases of development, at which a certain mechanism of regulations determining the growth character is kept up, is obvious. Open is the question about the possibility of determining moments of change in the tendencies of growth directly by the results of observations of quantitative characteristics of the growing system far removed from the start of the critical period. The solution of this very problem appears to be the object of the investigations whose results are presented in this book. One of the problems investigated was the simulation of moments when critical periods take place in the development of biological systems proceeding from quantitative information about the prehistory of their development.

Then, the proposed model was examined on systems different in level of integration and type of development. This examination has required the employment and comprehension of material from various sections of biology, geology and even astronomy, where the authors' competence, is naturally, insufficient. Therefore, certain items of work published in a number of papers were carried out in collaboration with different specialists: embryologist S. G. Vasetsky, population biologist A. V. Yablokov corresponding member of the USSR Academy of Sciences (who was also the editor of the Russian edition of the book published in 1982), paleontologists and geologists, academician B. S. Sokolov, and V. D. Nalivkin corresponding member of the USSR Academy of Sciences. The authors express their deep gratitude for their active and enthusiastic participation in examining the data from the point of view of nontraditional notions, and for their agreement to include the material from these common efforts in the present book.

The subject matter in this book was discussed at the general meeting of the General Biology Department of the USSR Academy of Sciences, at scientific boards and seminars of the Institute of Marine Biology, Far-East Geological Institute, Pacific Institute of Geography (Vladivostok), Khabarovsk Complex Research Institute, Institute of Volcanology (Petropavlovsk-on-Kamchatka) of the Far East Science Center, Institute of Developmental Biology, Institute of Evolutionary Morphology and Ecology and Institute of Oceanology (Moscow), Institute of Ecology of Plants and Animals (Sverdlovsk), Institute of Cytology, Zoological Institute, Institute of Evolutionary Physiology and Biochemistry of the USSR Academy of Sciences, the All-Union Oil Research Geological Prospecting Institute (Leningrad), and also with a number of scientists of whom we would like to mention academicians M. S. Gilyarov and V. A. Koptyug as well as G. T. Voronov. Astronomical data were discussed with academician A. A. Mikhailov, K. A. Tovastscherna, the director of the Pulkovo Astronomical Observatory and Prof. M. I. Pudovkin. The authors express their sincere appreciation to the mentioned scientists and to all who took part in the discussions for their attention, support, valuable critical comments and recommendations.

The book was translated into English by Irina Barsegova, Gennady Derkach, Tatyana Koznova, Aza Staviskaya and Tamara Voronovskaya. Great help in the preparation of the manuscript was rendered by Nina Kuryshkina. The authors express their grateful thanks to all of them.

1 The Problem of Critical Levels in the Development of Systems

The term development is understood as a directed, law-governed change of matter and consciousness; as a result of development, a new qualitative state of an object – its composition or structure – appears. "Two forms of development are distinguished between which a dialectic connection exists: an evolutionary form connected with gradual quantitative changes of an object, and a revolutionary one characterized by qualitative changes in the object's structure" (The Soviet Encyclopedic Dictionary 1980, p. 1109).

Such a comprehension of nature and the composition of developmental processes has been reached by human thought after a long struggle of ideas. But the elements of such views have been known for ages. Thus, in ancient Babylon, together with the idea of a world created by the divinity and its invariability, a doctrine of the cyclic development of nature was propagated, connected with changes in the position of heavenly bodies, as well as of "the great year" on the passage of which the same events had to be repeated on the Earth which had occurred at the beginning of the preceding "great year."

The ideas of stages in the development of the universe and the Earth connected with "the great year" were further elaborated by the ancient Greek philosophers Heraclitus and Empedocles. Engels (1878) notes that an original, naive but, as a matter of fact, right view of the world was characteristic of ancient Greek philosophy and first expressed clearly by Heraclitus: everything exists and at the same time does not exist, because everything passes, everything changes constantly, everything is in a constant process of appearance and disappearance. Engels further writes that dialectics was investigated in a more or less precise way by Aristotle and Hegel, and in spite of idealistic premises Hegel was the first who had given a comprehensive representation of its universal forms of development. Marx and Engels combined dialectics with materialistic views on nature and society and found a new conception of the world, dialectic materialism.

A study of the history of development of ideas in natural sciences, in particular in geology, paleontology and biology, will allow us to understand better the available material. Therefore, in order to discuss the development problem of natural systems along with the results in mathematics and the natural sciences, the authors consider also the history of the struggle of ideas in these branches of science.

In Sect. 1.1, questions of biological systems, hierarchical levels of their integration and the general system theory by Bertalanffy are discussed. In Sect. 1.2, a brief history of views on the correlation of evolution and leaps in development is given. Sect. 1.3 is devoted to the place occupied by the problem of critical levels in the study of the development of complex systems.

1.1 Biological Systems and Hierarchical Levels of Their Integration

In recent years, in philosophy and concrete sciences more and more attention has been given to the concept of systems. A system is considered to be a large number of elements connected with each other in different ways and forming a certain integrity, unity (Sadovsky 1976, p. 463). The system has a certain structure, that is, an interaction between, and influences on, the elements, where the elements, in their turn, have an influence on the system. The system interacts with other systems, including systems of a higher order, which are the external environment for it. Thus, systems have a hierarchical and multilevel character: they consist of elements which, as a rule, are systems themselves and, at the same time, are elements of systems of a higher order.

As far back as in the last century, in connection with the creation of the cellular theory, a presence of various levels of organization of living matter was recognized (Schleiden 1846; Virkhov 1865; Bernar 1878, and others). In the first half of the twentieth century, this idea was further developed both by philosophers (Brown 1926; Sellars 1926) and biologists. Different levels of organization of biological systems, their functioning and integration were discussed, as well as approaches to studies on different levels.

In lectures delivered in 1912–1913 Pavlov spoke of three floors on which research in physiology must be done, physiology being likened to a building: on its upper floor an analysis of the interrelations and the highest balancing of organism and environment is made, the middle floor is for the physiology of different organs and systems of organs which was most developed in those days and the lower floor is formed by the physiology of the cell which is "the bottom, the basic of life" (Pavlov 1952a, p. 65). In his other paper, Pavlov wrote about Heidenhain, a German physiologist, a representative of "the physiology that must replace our modern physiology of organs and that may be regarded as a harbinger of the last stage in the science of life, the physiology of the living molecule" (Pavlov 1949, p. 162).

Thus, Pavlov singles out four levels of physiology to be studied – intact organisms, organs, cells, and living molecules.

In the middle of the twentieth century, a number of biologists, sociologists and philosophers had worked out a theory of integrative levels of organization. The basic principles of this theory are formulated in a paper by Novikoff (1945), a biologist who stated that the conception of integrative levels of organization is a general description of the evolution of matter through successive and increasingly higher orders of complexity and integration. It considers the development of matter, beginning with cosmological changes resulting in the formation of the Earth and on to social changes in society, as uninterrupted because it never ends, and as interrupted because it passes through a series of different levels of organization. This conception does not reduce phenomena of a higher level to those of a lower one as in mechanism, and it does not describe phenomena of a higher level in vague, nonmaterialistic terms which replace comprehension as vitalism does (p. 209). Novikoff distinguishes the following levels of organization of living matter: cells, tissues, organs, systems of organs, organisms and populations.

It was Von Bertalanffy (1952) who played a great part in the development of ideas of the presence of spatial hierarchy of interrelated systems in living nature. Von Bertalanffy's (1949) classification of hierarchical orders of a "living level" was expressed briefly by Setrov (1971, p. 41) in the following series: field, plasm → elementary particles → atoms → molecules → aggregates of molecules, micelles → colloids → cells → tissues → organs → organisms → colonies, communities of living organisms.

Considerable interest in the levels problem has been shown by scientists working in various fields of biology: in zoology by Gilyarov (1954), Beklemishev (1964), Markevich (1968); in botany by Lavrenko (1965); in physiology by Bullock (1955); in medicine by Davidovsky (1958), Shumakov et al. (1971); in cytophysiology by Ushakov (1963), Alexandrov (1975); in biochemistry by Szent-Györgyi (1960), Poglazov (1970), Engelgart (1970); in cytology by de Robertis et al. (1970); in marine biology by Zenkevich (1967); in evolution biology by Schmalhausen (1961), Timofeev-Resovsky et al. (1977); in ecology by Dice (1952), Naumov (1964), Odum (1971), and others. They selected different series of structural, functional, organizational and integrative levels of "everything living", "life" or living matter, discussing their fields of interest in special detail.

Philosophers too pay no less attention to this field. Special monographs (e.g., Kremyansky 1969; Setrov 1971) and collected works (*The problem of integrity in modern biology,* 1968; *The problem of levels and systems in scientific cognition,* 1970; *The development of conception of structural levels in biology,* 1972) are issued, complex schemes are constructed by authors who make attempts at uniting all known levels into an integrated system (Vedenev et al. 1972, p. 67; Setrov 1972, p. 315), and discussions develop about a number of levels of living matter: initial and final levels, quantities and types of systemic hierarchies.

One of the most controversial questions is that of the lower level of life. Is it possible to speak of "physiology of the living molecule" as Pavlov (1949, p. 162) did, to single out molecular (Engelgart 1970), physico-chemical (Ovchinnikov 1980) and even submolecular (Szent-Györgyi 1960) biology and determine life as "self-realization of potential possibilities of electronic states of atoms" (Bernal 1969, p. 17)? An interesting press conference on this question was organized by the Soviet popular-science magazine "Nauka i zhizn" (Science and life; 1962 no. 4). It demonstrated a considerable diversity of viewpoints of soviet biologists on the question whether viruses and protein molecules are alive.

No less strong disagreements arose in the field of ecology. Some research was carried out at the interface of cytology and ecology, and was called cytoecology (Alexandrov 1952, 1975; Ushakov 1963; Biebl 1965; Zhirmunsky 1966) investigating cellular adaptations and cellular mechanisms of supercellular adaptations, which cannot be, in the opinion of some scientists, attributed to ecology (Zenkevich 1967; Gilyarov 1973) because ecology, from their point of view, must investigate only organismic and superorganismic adaptations.

According to Macfadyen (1963), the main object for ecology is populations, while other researchers make biocenoses, biogeocenoses (Sukachev 1957) or ecosystems (Tansley 1935) its cornerstone. Lastly, owing to the fact that attention was drawn to relations between man and the biosphere, numerous authors, mostly nonbiologists, began to speak and write about the ecology of the biosphere or "global ecology" (see, e.g., Budyko 1977).

Another group of controversial questions has appeared in connection with comprehending the plurality of ways in which life on the Earth develops. Are colonies of unicells multicellular organisms? What is an infusorian – a cell or an organism? Should populations and species be referred to different or unified "population-species level"?

Ushakov (1963), discussing the role of different levels in processes of adaptation and evolution, offered to distinguish the following 10 coordinative levels: molecular, organoid, cellular, organ-tissue, systems of organs, organismic, family-stock (reproductive), population, specific and interspecific (cenotic). The last of Ushakov's levels is given by Lavrenko (1965) as biocenotic. He has also specified a biostromatic level, corresponding to the totality of living matter on our planet.

Zavadsky (1966, p. 41) subjected Ushakov's (1963) and Naumov's (1964) schemes to criticism, considering their approach to the classification of levels to be formal, the series offered by them "miscellaneous" and "artificial", and their hierarchies characterized by the "coexistence of in principle heterogeneous systems – primary and universal (organism, species, biocenosis, biosphere) – with secondary and special (organ, tissue, family, stock, etc.)." He offered his own scheme: "Basic forms of organization of all living matter, main stages of their evolution, and structural components of each stage" (p. 42). This scheme is, however, unsuitable for use not only because of its extraordinary complexity in determining organismic and "population-species" stages of "primary systems", but also owing to the fact that really existing levels of organization of living matter available for investigation disappear in this system.

Basing on Vernadsky's (1978) notions of the simultaneous appearance of a complex group of forms (organisms which simultaneously were species, composed biocenoses and, in the aggregate, the biosphere), Zavadsky excludes from his hierarchy of "basic, primary forms of organization of everything living" the cell, the main brick of which the living organisms are built. Accusing biologists of "artificial" linear succession, Zavadsky involves himself in contradictions, not having taken into account the correlations of different hierarchical systems elaborated by Woodger (1937) and von Bertalanffy (1960).

According to von Bertalanffy, several hierarchical series of systems can be specified namely: hierarchy of division (for instance, in the reproduction of tissue cells or in protozoans); spatial hierarchy of a multicellular organism consisting of organs, tissues, cells and nuclei; genetic hierarchy (for instance, phylogenetic) – a correlation of organisms belonging to a species with organisms of various genera, families, etc.; physiological hierarchy of processes, an example of which is the functioning of the nervous system on different levels – intimate processes going on in nerve cells and fibers which form the basis of realization of reflective acts, which in their turn, are regulated by the activity of the central nervous system.

As Setrov (1966, 1972) mentions, spatial and genetic hierarchies should be declared to be most important for biology. After all, the clash of these very hierarchies results in contradictions when classifying biological systems.

As mentioned above, Zavadsky's scheme lacks a level of cells which he indicates as "a structural component" of the organism. It is indeed so that cells are structural components of a multicellular organism but, in this function too, they are equal members of the spatial hierarchy, the series that includes both cell and organism. Besides, cells of protozoans, microorganisms, sexual cells of organisms with external fertilization (many marine invertebrates), and zygotes perform a dual function – that of cell and of organism (Polyansky 1959, 1971). In particular, they are immediately faced with the effect of environmental factors and therefore, as Ushakov (1963) points out, are subjected to the selection effect.

Setrov (1972) states that a similar mistake is made by Zavadsky (1961, 1968) when he unites population and species into the unified "population-species level." To all appearances, such a unification is made by a number of researchers because sometimes a species can be represented by a single geographical population. In addition, in some cases, individuals of a population belonging to one species are isolated for a long time and seem to be not connected with each other. However, as Setrov mentions, the fact that the connection may not be only spatial but temporal as well is not taken into consideration. Populations separated by barriers for a long time maintain their ancestral features, characteristic of the species as a whole, which they had obtained from their mutual ancestors.

Ushakov's (1959) data confirm convincingly this viewpoint – he has found that the heat resistance of muscles (genotypical species feature) of the sea urchin *Strongylocentrotus intermedius* from populations of the coastal waters of the Kurile Islands, Sakhalin, and Primoriye, isolated from each other in the course of several thousand years, is the same.

Another example of genetic relationship between generations is given by the experiments of Zhirmunsky et al. (1967) who showed by methods of oxygen isotope paleothermometry that the upper living temperatures of fossil and recent molluscs of one species were the same.

A high stability of the population system of the gastropod *Littorina squalida* of the Busse Lagoon (Sakhalin Island) is expressed in the maintenance of the average gene frequency of shell coloration during 2000 generations (Kalabushkin 1976).

Setrov (1971) comes to the conclusion that denial of the fact that a construction of "linear" hierarchical series is legitimate cannot be considered right, because these hierarchies really, do exist. The scheme of organizational levels offered by Setrov (1972, p. 315) seems to be the most successful of all "comprehensive" schemes which we know. It combines living matter and inorganic nature, takes into account the existence of various kinds of hierarchies and the dual nature of cells and populations, which are at the same time systems of different hierarchies. Naturally, it can be subjected to criticism by different specialists as to names and numbers of the concrete levels, but this is due first of all to the inadequate elaboration of a number of sections of the natural sciences.

It should be mentioned that some authors (Zavadsky 1966) offer a comparatively small number of "main" levels. In Odum (1971), these are the six "main levels of life organization": gene, cell, organ, organism, population, community[1]. In *The Great Soviet Encyclopedia* (Astaurov et al. 1970), other six levels

[1] It is worth mentioning that the first member of this series belongs, in accordance with von Bertalanffy's classification, to genetic hierarchy.

are given: molecular, cellular, organismic, population-species, biocenotic, and biospheric.

In Sect. 5.3 size ranges of different organismic structures are considered, which must be used when constructing a new scheme of structure-functional levels of the organization of living matter, proceeding from both the linear sizes of different structures and their functional significance.

A development of ideas on the organization of matter, its structural nature expressed in the existence of a hierarchical system comprising interrelated systems of different integrational levels has resulted in a quest for general and special laws characterizing these systems.

To this problem are devoted Bogdanov's (1912, 1925–1929) works in which a "general organizational science", tectology, applicable to any phenomenon, has been developed, with mathematical simulation as its basic principle. Tectology considers phenomena from the point of view of continuous processes of organization and disorganization and the discovery of mechanisms regulating these processes. A live protoplasm, Bogdanov (1925–1929) believes, is characterized by a movable equilibrium regulated by a specific "biregulator".

Analogous ideas were later developed by Bauer (1935) who elaborated a theory of "stable nonequilibrium," and by Nasonov and Alexandrov (1940) who propounded a conception of movable equilibrium of the live protoplasm, changes of denaturation and renativation in cellular proteins forming its basis.

Bauer (1935) supposed that the task of theoretical biology was to unite and express in the form of one or several laws what was typical of all living systems and characteristic of them only (p. 22). Von Bertalanffy (1952, 1960, 1968), author of the general system theory, sets biology a similar task. He believes that there are general principles governing the existence of any system regardless of the nature of its composing elements and the relations between them. These principles may be determined in terms of mathematics. The discovery of these principles and mathematical patterns is the basic idea of the general system theory. They will make it possible to put such main notions of biology as integrity, dynamics and organization of systems on an exact scientific basis.

1.2 On the Correlation of Evolution and Leaps in Development

As far back as in ancient times, together with views about the invariability of the world, notions of its development, gradual or uneven, were disseminated. A revival of materialistic philosophy in the seventeenth and eighteenth centuries resulted in the realization that it was necessary to develop sciences by taking experimental knowledge as a basis (F. Bacon), which did much for promoting the development of the natural sciences. But the domination of theological dogmas over the minds of naturalists left its mark upon ideas developed by them. Late in the eighteenth and early in the nineteenth century, three basic directions in the explanation of the principal laws of development of the Earth and of animate nature were successively formed, namely catastrophism, uniformism, and evolutionism (Wawell 1869; Huxley 1883; Ravikovich 1969; Zavadsky and Kolchinsky 1977, and others).

Though ideas of catastrophism were expressed in works by some naturalists of the eighteenth century, the most vivid reasons for it were given by Cuvier in 1812 in his work *Talks about revolutions on the surface of the globe* (Cuvier 1937). Cuvier propounded a hypothesis of periodically repeated catastrophes of a nonglobal character after which, from the surviving organisms, new communities originated. Twenty years later Lyell (1830–1833) published a book *Principles of geology* that took uniformism as a basis. Lyell furnished proofs of the gradual transformation of the Earth's surface and life conditions on it. Lyell's views influenced very much the development of the natural sciences. His theory led directly to the doctrine of a gradual transformation of organisms.

The notion of the evolution of living beings from simple forms to increasingly complex ones under the influence of environmental change was introduced by works of a number of naturalists, among whom Lamarck was the most remarkable scientist. However, owing to the fact that his theory was contradictory in the explanation of evolutionary causes and insufficiently substantiated, it had no effect on the development of science of his times. And only the publication of *The origin of species* by Darwin (1859) brought about the triumph of the theory of evolution in biology.

However, in the twentieth century, these theories which have been almost rejected already, found, based on new principles, new supporters. Just when A. Severtsov, Schmalhausen, Dobzhansky, Mayer, Timofeev-Resovsky, Simpson, J. Huxley and many other representatives of the modern "synthetic theory of evolution" affirmed that all evolutionary change in the past was caused by factors functioning nowadays as well, and so available to experimental examination (Dobzhansky 1937, p. 4) – these scientists belonging according to Zavadsky and Kolchinsky's (1977) classification to the "neouniformists" – there appeared a great many of supporters of "autogenetic", "ecogenetic", "cyclic" and "acyclic", "neocatastrophic" theories and their varieties, as well as of "variaformism", on the scene. According to the latter the possibility of different forms of change in the factors and causes of evolution is confirmed (Zavadsky and Kolchinsky 1977, p. 49).

As it is not possible for us to consider and discuss all these trends, we shall only mention that very many of the "neocatastrophists" are to be found among geologists and paleontologists. Thus, in the view of paleontologist Sobolev (1924), the history of the organic world is composed of small and large biogenetic cycles including periods of evolutionary development, interrupted by revolutionary periods. He believes that radical changes in fauna occurred on the boundaries between the Proterozoic and Paleozoic, Devonian and Carboniferous, Permian and Triassic, Mesozoic and Cainozoic, and coincided in time with the beginning of cycles of orogenetic processes. He also supposes that they were determined by cosmic cycles.

Schindewolf (1954, 1963) also pays attention to sharp changes in fauna composition on the boundaries of areas and periods of geological history and singles out cycles of several phases where the extinction rate and the rate of appearance of new forms are different. In Schindewolf's opinion, the role of selection is the elimination of types whose existence is impossible in the given conditions. The increase in evolutionary rates at certain phases Schindewolf connected with supposed changes in cosmic and solar radiation. Analogous notions were developed by Krasovsky and Shklovsky (1957).

Lichkov (1945) stated that a revolution in the organic world on an universal scale happened every time at periods of formation of folds in the Earth's crust (p. 169).

Sushkin (1922), Yakovlev (1922), Krasilov (1973) and some other scientists developed the idea of climatic changes having a decisive influence on the evolution of life. The first two authors also explained the change of taxonomic groups not by the supersession of groups (as Darwin did) but by the occupation of place which becomes vacant as a result of extinction due to unfavorable conditions.

A number of authors connected critical ages in the development of life with a cyclic strengthening of volcanism (e.g. Pavlov 1924), change in solar light intensity (Golenkin 1959), orogenesis (Kristofovich 1957), and intensification of the migration of radioactive elements to the Earth's surface (Ivanova 1955).

This list of views could be considerably enlarged but we believe that those mentioned above are sufficient for concluding that the question of the correlation of evolutionary and revolutionary phases in the development of the Earth and the organic world is by no means solved. To wind up by presenting extracts from books by Grant (1977) and Nalivkin (1980).

It has been long known – wrote Grant – that the geological history of life is noted for having episodes of mass extinction separated by prolonged periods of gradual evolutionary change. The paleontological chronicle makes us presume something which lies between Cuvier's catastrophism and Lyell's strict uniformism, that is, a state of episodic evolution. In history, episodes of mass extinction of one group appears followed periodically by the development and adaptive radiation of new groups (Newell 1967). An important feature of the episodes of extinction is the high rate of extinction observed in absolutely unrelated groups at approximately one and the same time. Thus, in ammonoids and reptiles, quick extinction is observed in the Late Permian, Late Triassic, and Late Cretaceous. The majority of these extinction periods can be correlated with periods of elevation of continents and the establishment of continental climate. Such a correlation makes it possible to believe that the perturbations that happened in the environment starting complex chain reactions resulted, after all, in extinctions on a large scale (Newell 1967; Axelrod 1974). The episodes of mass extinction are followed by the formation and development of new groups. The necessary condition for a new group to appear is the presence of corresponding ecological possibilities which resulted from mass extinction[2] (p. 328–329).

Closely similar views of the question have been developed by Nalivkin. He states that "the Earth is part of the Universe and so great events that happened in the solar system and the Galaxy influenced its development and structure. In the Earth's history, there existed periods of calm evolutionary development and active revolutionary reconstruction... . Evolutionary periods are longer than revolutionary ones. In revolutionary periods, the continents usually rose and the sea receded, i.e. maximal regressions occurred" (1980 p. 6–7).

---

[2] The latter conclusion, as we see, coincides with the notions stated before by Sushkin (1922) and Yakovlev (1922).

The struggle of ideas in respect of development of the Earth and the organic world had a dominant influence on the views of naturalists. The viewpoint of Leibniz that "nature does not make leaps", taken up by Lamarck, Lyell and other scientists, was accepted by a number of leading mathematicians and resulted in opinions of a type given by Lorents (1897) in defence of the necessity to use only uninterrupted models. He wrote that even if one would accept the possibility of interrupted change in nature, it is still possible, when investigating such a phenomenon, to replace it with the process of uninterrupted change and then determine how an increase in the change rate will be reflected in the results obtained. Just such views have resulted in the creation of the main features of the modern mathematical apparatus, e.g., equations of mathematical physics. Such approaches to modeling, while describing the developmental aspects in the evolutionary process, do not make it possible to consider trajectories of development which include transitions through points of qualitative changes.

A decisive influence on the realization of the principal role of leaps in the development of biological systems is to be found in an introduction by de Freeze of the conception of mutations, called by Schrödinger (1945) "quantum leaps", in the gene molecule, which are spasmodic changes occurring without intermediate states.

Later on, the idea of quantizing the physical processes led to the creation of quantum physics. Vavilov (1943) wrote: "New physics, in some points, has rejected the idea of continuity; the idea of atomization, leaps and interruptions, has deeply penetrated into modern science. Mass, electrical charge, energy, action ... are atomized" (p. 137).

Schrödinger (1945) believed that the time of waiting of mutation t with the energy threshold W was determined by the function

$$t = \tau \exp W/kT \ ,$$

where τ is the period of fluctuations, k is Boltzmann's constant, and T is temperature.

In accordance with experimental data given by Popova (1971) for the pike, $t = 0.3\, e^{1.4\tau^*}$. The value of τ^* is the time of delay, owing to the circumstance that succession of mutations is connected with characteristics of the system's "memory" about its prehistory. It is a matter of principled influence of "memory" of preceding states in time on the current system characteristics.

Similarly, an influence of spatial "memory" is repeatedly noted, as the effect of neighboring elements on the development of the given one (Gurvich 1977; Dewcar 1978; Svetlov 1978, and others).

At the same time, basic models used nowadays for the description of the development of biological systems include, as a rule, only a dependence of the system's growth rate on its size and age at the given moment, and do not take into account the influence of prehistory (Mina and Klevezal 1976).

The practice of modeling developmental processes shows that models of a certain kind prove to be right only within the quite certain range of changes in the object's parameters and in the time of its development. This is, as a rule, connected with the variable character of internal and external conditions of the object's development.

It is, for instance, established that at constant internal and external developmental conditions, the enlarged reproduction occurs in accordance with the law of complex percentages or corresponds to the exponential law with a constant value of relative increments. Once these conditions have changed, beginning with a certain moment in time, adaptive changes in the system's structure and the character of its regulations will be required to balance the new developmental conditions (Ashbi 1962).

Thus, here the dialectical law of transition from quantitative to qualitative changes is reflected; it must be expressed in descriptions of the structure of economic processes used for modeling. In this connection, an elaboration of the following models is necessary:

– models describing a change in the object's characteristics based on the certain principle of its action (evolutionary models);
– models intended for determining terms of use of the certain principle of action, including the moments of transition of the new principle to action.

1.3 Critical Levels in the Development of Complex Systems

Modeling of critical levels in the development of systems, which are understood as moments of radical qualitative change, is of considerable theoretical and practical interest. Critical levels determine the ranges within which the system has different morphofunctional structures or principles of regulation, and transition through these levels is connected with corresponding structural and regulatory changes.

Between successive critical levels the system retains its qualitative properties; it must be characterized here by a low sensibility to external and internal changes in developmental conditions (high resistance). For such ranges one may expect a below-the-average reaction to external effects, and with these models description of such developmental stages can be extremely simple (with a small number of parameters) and without loss of accuracy in the description.

Ranges of increased sensibility to external effects are recorded in experimental embryology (Svetlov 1960) and called critical. At these periods, an increased rejection of defective embryos is observed through the action of the most varied irritants. Such is, for example, the gastrulation stage, the rejection at this stage being comparable with that at birth.

Between gastrulation and birth, there is a number of critical periods but they are considerably less significant. Hence, there arise problems of classification of critical level types, the establishment of their ranks of significance, and the determination of the nature of changes typical of each rank.

In the organization of biological systems the presence of critical levels as regards populations is observed as stable amplitudes in population abundance fluctuations (the "waves of life" of Chetverikov 1905), the upper and lower boundaries being kept up at an approximately steady level during long intervals of time (many dozens and even hundreds of years). Discovery of the theoretical limits in the development of biological systems of different hierarchical levels, and the establishment of interrelations between the levels and their general quantitative formation patterns, are very important for the development theory of biological systems.

In connection with determining critical levels practical results are necessary for controlling system development. Commercial exploitation of a population up to a level below the certain critical abundance results in its breakup. To determine permissible exploitation limits is the purpose of research. One and the same pressure on an organism can be damaging at one phase of its development, therapeutical at another, or it may exercise no visible influence at all. How to classify developmental phases by their sensibility to controlling effects? Questions of this kind, specific for systems of each hierarchical level, can be formulated together with rather detailed and concrete stating of problems. But common to these is the necessity to determine critical levels in system development on different hierarchical levels and to work out the peculiarities of its functioning in detail within the limits of each critical range.

One of the purposes of the present book is to investigate the question of general approaches to the modeling of critical levels of systems development, interrelations between the characteristics of critical levels, and formation of critical developmental ranges for consecutive hierarchical levels.

Another problem concerns examining the effectiveness of models worked out by us on the examples connected with the development of natural systems: astrophysical, geological, and biological.

In the final section of the book, the possibility of extending patterns found onto other systems is considered and a law of critical levels of systems development is formulated.

2 Model of Systems Development

Studies of natural systems show that they may be regarded as multilevel systems which are characterized by strong interrelations between their different structural levels, among the elements of each level, and between the systems themselves and their environment. The following facts suggest how complex are the systems to be considered. A human organism comprises a hierarchy of structural levels whose size lies at least in the range from 1.7 to 10^{-9} m (Zhirmunsky and Kuzmin 1982).

The similarity of orders of magnitude is displayed by the time characteristics of the organism's cycles, which cover a range from 100 years (an approximate duration of life) to thermal radiation rhythms of 8–10 mkm ($3 \cdot 10^{13}$ Hz; Gulyaev and Godik 1984). This range is fully occupied by the rhythms of various organic systems and the rhythms of interactions between them.

The above mentioned system features are characteristic not only of biological objects. These are the intrinsic features of any developing system. We make use of examples from biology to comprehend the most complex of possible phenomena and objects.

In what manner do such diverse systems act on another? How are common activity rhythms set? How are models to be built, which would allow one to realize in a unified manner how systems function and act reciprocally on each other and the environment? And, on the whole, are there any quantitative relationships conforming to the laws of nature and describing the development and functioning of such complex systems?

To answer the questions we need to build a development model which will describe evolutionary intervals and leaps of development separating the qualitatively uniform intervals on different hierarchical levels. To describe the development process of any hierarchical level we formulate an associated development equation based on the concepts of chain processes and the effects of the prehistory of development. The aim is to formulate a development equation that permits generalizing the known models and in an aggregate of the conventional model for processing experimental data (the Pearson model as the generalization of density functions in mathematical statistics, the hypergeometric equation as the generalization of special functions in the theory of oscillations, and so on). The development equation is an extension of the wide class of conventional models. The processes on different hierarchical levels are described by the development model, which is made up of the development equations obtained by taking their logarithm or inverse logarithm, ensuring the soundness of the model for each level.

This section covers the issues concerning the formulation of the development equation, its connection with conventional models, and setting a scale of arguments for defining process dynamics on each hierarchical level.

2.1 Developmental Equation

The forming of the development equation is based on the following axioms:

1) A development process is defined by some system parameter (called a base variable) which is a numeric characteristic of the branching process, or chain reaction.
2) A development process is significantly dependent on its prehistory.

The first of the two axioms states that the rate of change of a base variable is proportional to its value, which in a simple case results in an exponential relation. The second axiom says that the rate of change of a base variable is affected by its value in the past.

In this section we will discuss the matter of the axioms, derive a development equation based on them, and examine the sense of the equation considering its relation to the wide classes of conventional models.

Let us consider the basic quantitative relationships of a system development process by analyzing the dynamics of system growth.

To begin with, a development proceeds as a birth-and-death process. A chain mechanism forms the basis of the process (in chemistry and nuclear physics, the process is called a chain reaction). An increase in the number of system elements and in the system size is then determined by processes of division. Two cells develop from one cell in its intrinsic time, either of the cells, in its turn, produces two cells, and so on. The number of elements undergoing the chain reaction obeys an increasing geometric progression and is described by the equation of exponential growth in which the rate of system growth, $\dot{x}$, is directly proportional to the current system size, x,

$$\dot{x} = kx \tag{1}$$

where k stands for the growth constant. Integration of the equation leads to

$$\ln x = kt + \ln C \,, \qquad x = x_0 e^{kt}$$

whence a process obeying the law of exponential growth is linear if it is plotted on semi-logarithmic coordinates.

The exponential relationship is shown to describe the entire process of system development under the stationary internal and external conditions. This was noted by Schmalhausen (1935) who cited the rod bacterium *Bacillus megaterium* as an example of the system which develops under stationary internal conditions. The bacterium is fed through its surface, and the area of its surface is directly proportional to bacterium length owing to its relatively small thickness, while bacterium “volume” should be fed, which is also proportional to the length in this

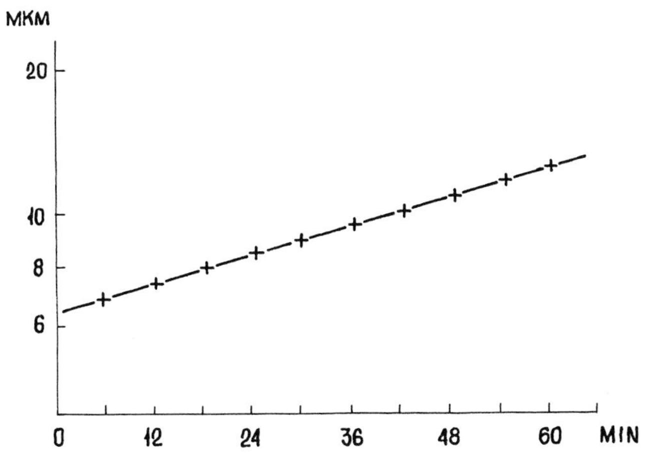

Fig. 3. Growth of rod bacterium (*Bacillus megaterium*). *Abscissa* age; *Ordinate* length; *MKM* (μm) (logarithmic scale) (Data from Schmalhausen 1984, V. 2, p. 114)

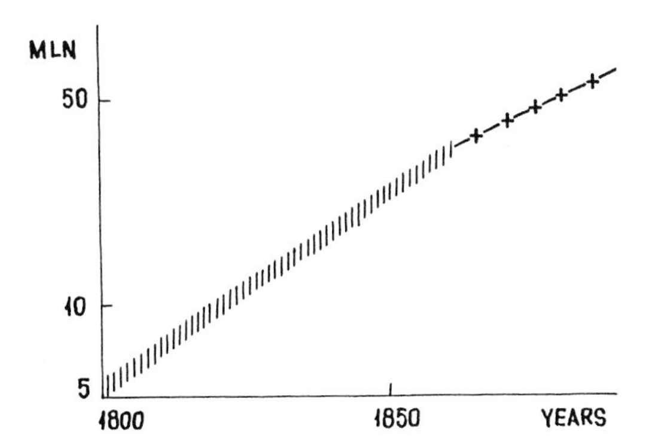

Fig. 4. Growth of the population in the USA from 1800 through 1860. *Abscissa* age; *Ordinate* population in millions (Data from Mendelson 1959)

case. It means that a volume fed through a unit surface does not change during the growth of the rod bacterium and, as a result, its growth proves exponential (Fig. 3).

The nonstationariness of the internal and external development conditions suggests that an exponential growth with constant rate ceases in certain states of the developing system and the medium in which it develops. Nevertheless, the periods, or modes, of exponential growth prove sufficiently long, not allowing one to ignore them. For instance, the growth of population in the USA remained exponential for 60 years, from 1800 through 1860 (Fig. 4).

Thus, the modes of exponential growth are associated not only with development under stationary conditions, but also take place when the conditions vary within some limits defined by certain critical quantities. Finding these critical quantities is, in our opinion, one of the fundamental problems of development modeling because this allows one to set applicability bounds to data concerning the formed trend in growth. Brody (1927, 1945) used partly exponential growth modes for modeling the development of biological systems and he determined, for each case, bounds to the applicability of the exponential relationship based on morphofunctional attributes. Schmalhausen (1935) noted that growth processes began with an exponential phase. But he believed that the exponential growth then gave way to a parabolic one.

The parabolic growth takes place when growth constants decrease from one exponential mode to another, so that they can be approximated by a hyperbolic dependence, $k(t) = B/t$. In other words, here the growth constant proves inversely proportional to an age. Introduction of the dependence into Eq. (1) leads to

$$\dot{x} = \frac{B}{t} x \tag{2}$$

or, after integration,

$$\ln x = B \cdot \ln t + \ln C$$

from which

$$x = A t^B . \tag{3}$$

In this case a growth curve is a graph of power function. This is exemplified by the growth dynamics of ball-shaped bacterium which is fed through its surface, while its volume should be fed. The area of its surface, S, is proportional to the square of its diameter, d: $S \propto d^2$, and its volume to the cube of the diameter: $V \propto d^3$. It means that a volume $V/S \propto d$ is fed through a unit surface, i.e., the volume fed through a unit surface increases with bacterium growth; as a result, the growth rate decreases (Fig. 5). As seen from Fig. 5, a relative increase $\dot{x}/x$ turns out to be the hyperbolic function of an age, and the system growth proceeds in

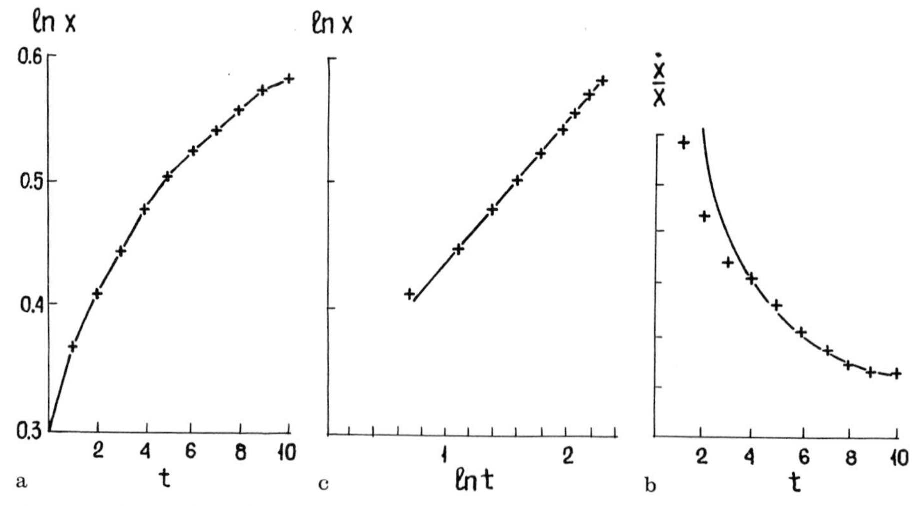

Fig. 5a–c. Growth dynamics of the ball-shaped bacterium Micrococcus; **a** growth in diameter; *ln* μm; **b** relative growth (hyperbola B/t is shown by the *solid line*); **c** Growth in diameter. Logarithmic scale for both axes. Time in tenths of an hour is plotted on the *abscissa* (Data from Schmalhausen 1984, V. 2, p. 115)

a linear manner provided logarithmic scales are chosen for both axes (Fig. 5c). This indicates that some mechanisms reveal themselves in system development processes, which cause a discrete change in the rates of exponential growth at certain development moments. The model described by Eq. (1) contains no information as to what factors give rise to the changes. That is why Brody (1945) endeavored to use information on morphofunctional transformation in the organism as the indication of a change in the growth rate. This indicates that the model of exponential growth describing a development at certain stages does not take into account some significant factors. What are these factors?

Notice that Eq. (1) describes a development process as a Markov one, i.e., a growth rate is dependent only on a current system state, but it is independent of its prehistory at all. At the same time, the interactions between system elements and the response of a system to external actions, in principle, possess lag properties. The necessity of taking into account the lag properties of systems has been discussed in literature for a long time. In the early years of the thirteenth century Fibonacci (Leonardo of Pisa) solved a problem concerned with an increase in the number of rabbits. Rabbits breed every month and have the first litter in two months after birth. This results in a Fibonacci sequence given by a recurrent relationship $x^{k+1} = x^k + x^{k-1}$ where x^k is the number of rabbits in the kth month. Hence, it is essential to consider not only a current quantity, but also a quantity in the previous month.

The fact that prehistory has a significant influence on system development was noted by Pirson (1911). He wrote (p. 637) that the instantaneous transfer of effect of body P, which suddenly begins to move, on remote body Q seems unlikely; it takes some time for a change in the position of P to tell on body Q. On developing the mathematical theory of the struggle for existence Volterra (1976) took account of the influence of aftereffects, or memory, on a current system state. When developing the theory of intracellular process dynamics Goodwin (1966) pointed out that the transition from differential equations and integral invariants to functional equatitons and invariant measures would made the theory more powerful, allowing one to consider heterogeneous systems instead of homogeneous ones, as well as systems with delay and hysteresis effects (p. 32).

At present, the trend to take into account the influence of spatial and time prehistories on a current system state, becomes so apparent that the development of continuum mechanics is beginning. Continuum mechanics is based on the postulate of essential influence of prehistories on process development. As Truesdell (1975) noted, the prehistories play an important role in mechanics, for the future is actually defined by the present and the past. In 1959, Noll set up a set of axioms which resulted in continuum mechanics to be formalized on a new basis. The first axiom (called a determinism principle) states that the stressed state in the configuration of a body-point, X, at the moment, T, is defined by the prehistory of body motion up to the moment T. Thus, the past and present configurations of a body determine the stress field acting upon the body in the present configuration (Truesdell 1975).

Similar observations are made on the nature of nonlinear effects in radioelectronics – the nonlinearity of a medium can also be attributable to the lag properties of the matter which manifest themselves in the fact that medium parameters vary under the action of a propagated electromagnetic field rather with a finite velocity than instantaneously. As a result, the response of the medium to the acting field is delay for some finite time which is dependent on a field magnitude.

If we consider the lag properties of a development process in Eq. (1), i.e., in accordance with Noll's axiom, the growth rate is considered to be in direct proportion to the system size at the moment shifted from the present time by a characteristic delay time, τ, and we obtain

$$\dot{x}(t) = kx(t-\tau) \ . \tag{4}$$

To solve the equation an initial function should be given, i.e., we should have a trajectory $x(t) = \varphi(t)$ for $t \in (t-\tau, t_0)$. A more general case can be obtained if parameters k and τ are considered to be time functions.

$$\dot{x}(t) = k(t)x[t-\tau(t)] \ . \tag{5}$$

Equation (5) will be referred to as a development equation.

In what follows, we discuss the relationship between the development equation (5) and different conventional models.

2.2 The Communication of the Developmental Equation with Main Models of Experimental Data Processing

In trying to build a system development model we should demand, as a test for its generality, that the development model should include, as particular cases, universal models that are used for processing experimental data in various research areas. There are, in principle, a few models of the kind. These are models of growth, fluctuation, various models of probability theory and mathematical statistics, information theory, catastrophe theory. Models of growth and fluctuations are intended to describe the dynamics of processes. Models of probability theory and mathematical statistics are designed to analyze data of a statistical section belonging to a particular moment, as well as dynamics (theory of random processes). Models of information theory treat the characteristics of processes of message transformation and transmission.

The catastrophe theory, which has been advanced a short time ago, studies phase transitions in system development. We successively examine the relationship of these classes of models with the development equation.

As stated above, exponential growth plays an essential role at an individual stage of system development processes.

Equation (4) is noted for the fact that the exponential growth is its particular case. In order to obtain an approximate solution to the equation, a function of delayed argument $x(t-\tau)$ is expanded by its Taylor series expansion with only two terms retained. In this case additional terms must not be taken so as to avoid an oscillation behavior not inherent in an original process (Elsgolts and Norkin 1971).

$$\dot{x}(t-\tau) = x(t) - \tau\dot{x}(t) \ . \tag{6}$$

Introduction of (6) into Eq. (4) leads to

$$\dot{x}(t) = \frac{k}{1 + k\tau} x(t). \tag{7}$$

It is obvious that Eq. (7) coincides with Eq. (1) of exponential growth with constant relative growth or growth rate, provided k and τ are constants.

The main trends in the growth of biological systems have been studied in many works (Huxley 1932; Schmalhausen 1935, 1984; Brody 1945; Medawar 1945; von Bertalanffy 1952; Vinberg 1979, and so on). The studies were summarized by Medawar (1945). He wrote that only one fundamental statement could be made concerning the relationship between the size and age of an organism:

$$\dot{x} = k(t)x\ , \tag{8}$$

where k(t) is such a positive number that k(t) decreases and $\dot{k}(t)$ vanishes when t increases. Besides, a particular case (k = const) is singled out; it corresponds to exponential growth.

The comparison of development equation (5) with Medawar's model (8) shows that the equation and model coincide with each other, provided there is no delay in (5), i.e., when $\tau = 0$.

Schmalhausen (1935, 1984) took Eq. (8) as a basis for his study, which led him to the model of parabolic type. In this case the right-hand side of Eq. (8) represents a specific growth rate, i.e., a gain per unit size of a system for unit time. Schmalhausen analyzed trends in specific growth rates as a function of the age of a developing system. Based on experimental data he found that the quantity was inversely proportional to the age, i.e., $k(t) = k/t$. This approach to the building of a growth model seems promising because it allows one to determine the structure of function k(t) by examining changes in specific growth rates and then to obtain a growth curve using the direct integration of Eq. (8).

Thus, Medawar's equation (8) is a particular case of development equation (5).

A potential function $V = V(x_1, x_2, \ldots, x_n)$ is used in analytical mechanics to describe trajectories of dynamic systems. A process equation then takes on the form

$$\dot{x} = \text{grad}\, V(x) \tag{9}$$

where x is the state of a system. The direction of the gradient of a function coincides with the direction of maximum increase of the function. This reminds one of extremum principles of physics. At the same time, a lot of works deal with the study of physiological gradients as determining factors of organism regulation and development (Child 1929; Gurvich 1977). Longitudinal gradients have been experimentally obtained for exchange phenomena at early stages of the development of backboned animals. The gradients direct organism growth and manifest themselves in the phenomena that integrate an organism as whole.

Suppose that a system develops in an anisotropic field. The spatial nonuniformity of the field causes a field gradient. System development is believed to take place in the direction of the field gradient. The process equation is then written in the form of Eq. (9). The structure of grad V(x) is defined by the memory or nonlocal properties of the system (by the effects of neighboring elements on a given one, i.e., by spatial memory). A modeling problem, in principle, consists in building standard functions to describe grad V(x).

The introduction of concepts of field and its gradient (with the structure of grad V(x) specified for a given process) permits modeling a transition from quantitative to radical qualitative changes. At present, the results of the catastrophe theory are applied in exactly the same way (Poston and Stuart 1978, Gilmor 1984). The term "catastrophe theory" is borrowed from Cuvier in connection with the study of structural stability and its application to the classification of abrupt transitions in developing systems. The founder of the modern catastrophe theory Thom (1968) pointed out that two sources formed the basis of the theory. The first of them is the topological and analytical study of structural stability. The object of the study is to ascertain if a given function, when perturbed, retains its structure. The second source is works on embryology. Here a development process is divided into fields of structural stability or determinism which are separated by zones where the process is indeterminate and structurally unstable. The similar features are characteristic of processes considered in geometric optics and fluid mechanics (Poston and Stuart 1978). This gives reasons to believe that the theory (it is closely bound up with the study of general properties of structural stability of models) has a universal character.

Processes described by Eq. (9) are treated in the catastrophe theory. Thom (1968) has shown that, for dynamic systems which are described by differential equations with smooth functions on their right-hand sides and whose output characteristics experience jumps, the trajectories of these abrupt changes can be built using geometry techniques. The number of types of jumps depends on the number of system parameters. The catastrophe theory deals with a system whose potential function is dependent on the collection of parameters $C_1, C_2, \ldots, C_k$. The system is supposed to be described by a smooth function V_c at every value of $C = (C_1, C_2, \ldots, C_k)$. The equilibrium position of the system may vary with parameters. The parameters themselves undergo slow changes. Then the system is always in equilibrium position, which corresponds to the minimum of function V_c. If a current equilibrium position vanishes because of a change of parameters C, the system passes to one of different equilibrium positions. The basic result obtained by Thom (1968) is represented by the following theorem. "For a system described by a smooth function with no more than four parameters and having any number of variables, there exist only seven possible types of local geometrical structures for stable catastrophe sets."

The strong point of the catastrophe theory is the correlation of catastrophes with real processes in comparison with different types of grad V(x). The weak spot in the theory consists in using models of the Markov type. After all, a catastrophe, as a rule, results from an interaction with different hierarchy levels, neighboring elements and systems, as well the characteristics of internal development processes. An interaction with different elements, the effect of development prehistory, a change in the environment, which in turn affect lower hierarchy levels, result in a change of development pattern, in jump, in catastrophe. The fact that parameters enter into a model in great numbers may appear to be due to the factors which are not directly taken into account, but are essential to the description of a process. Hence it comes to mind that models of a non-Markov type will be more attractive for catastrophe modeling.

In Eq. (9) the rate of change in the system state can be linearly approximated by difference

$$\frac{x(t)-x(t-\tau)}{\tau}$$

or by an increase in size taken with respect to the characteristic process time τ. In general, the rate of change of the system size is proportional to a change in size for characteristic delay time.

$$\dot{x} = B[x(t)-x(t-\tau)] \ . \tag{10}$$

Here the first term describes the effects of the growth mechanism, and the second one the effects of the die-off mechanism. In this case, the value of the function gradient is defined by the conflict of two basic trends. One of the trends is associated with the processes of element integration and reproduction (assimilation processes), and the second one with the processes of element disintegration and die-off (dissimilation processes).

By making the substitution of variable

$$\xi = x \exp(-Bt)$$

Eq. (10) is reduced to

$$\dot{\xi} = -B \exp(-B\tau) \cdot \xi(t-\tau) \ . \tag{11}$$

When $-B \exp(-B\tau) = k(t)$, Eq. (11) is equivalent to development equation (5). Thereby, the representation of the gradient of a potential function by a function of delayed argument gives Eq. (10), which is equivalent to development equation (5), up to coordinate transformation. Hence, development equation (5) can be expected to include results which are obtained in the catastrophe theory.

Equation (11) is noticeable for the fact that it represents a generalization of the Pearson models, whose integral is density functions which are frequently used in the probability theory and mathematical statistics for processing experimental data. The Pearson equation looks like (Kramer 1975).

$$\dot{x} = \frac{a+bt}{a_2 t^2 + a_1 t + a_0} x \ . \tag{12}$$

Solving the Pearson equation yields Gaussian, exponential or Poisson distribution depending on the parameters of Eq. (12) and so on. In principle, the Pearson equation (12) gives specific expression to function $k(t)$ in Medawar's equation (8).

We will show that the Pearson equation (12) is a particular case of Eq. (11) with a delayed argument. The expansion of the right-hand side of Eq. (11) in a truncated Taylor series with two terms (as stated above, the number of expansion terms is restricted so as to avoid an oscillation behavior which is not inherent in a process) yields the following relation

$$\dot{\xi}(t-\tau) = \xi(t) - \tau\dot{\xi}(t) \ .$$

Its introduction into Eq. (11) leads to

$$\dot{\xi}(t) = \frac{-B\exp(-B\tau)}{1-B\tau\exp(-B\tau)}\xi(t) \ .$$

Using the linear approximation of the exponential function in the right-hand side of the expression, we obtain

$$\dot{\xi}(t) = \frac{-B(1-B\tau)}{1-B\tau(1-B\tau)}\xi(t) = \frac{-B+B^2\tau}{1-B\tau+B^2\tau^2}\xi(t)$$

or for $\tau = At$

$$\dot{\xi}(t) = \frac{-B+B^2At}{1-BAt+B^2A^2t^2}\xi(t) \ .$$

Note that Eq. (11) coincides with the Pearson equation (12) for $\tau = At$. Equation (11) represents the model of a birth-and-death process, as it is a consequence of Eq. (10). Density functions in probability theory are, as a rule, unimodal functions, characterizing the conflict between an increase and a decrease trend.

The Pearson equation (12) is the generalized model of the family of density functions used in mathematical statistics. The coefficients of the development equation (5), as well its particular case (11), which is the generalization of the Pearson equation, can be physically interpreted as measures of system memory, while the coefficients of the Pearson equation are empiric factors.

The wide application of techniques of mathematical statistics to the processing of experimental data allows us to expect that the development equation (5), as an extension of the Pearson equation, will also describe general properties of a variety of systems.

The relationship between density functions and the development equation can be shown in another way. By the ergodic theorem, time means are equal to phase-space means. A statistical section contains information concerning the dynamic properties of a system, and a time series includes information on the structure of data describing a population. The relationships may be revealed in an analytical form using the model of a dynamic process of a sufficiently general type

$$(a_m D^m + a_{m-1} D^{m-1} + \ldots + a_0)x(t) + (b_p D^p + b_{p-1} D^{p-1} + \ldots + b_0)x(t-\tau) = 0$$

where a_i, $i = 0, 1, \ldots, m$, and b_j, $j = 0, 1, \ldots, p$, are the constant coefficients, τ the constant delay and D the differentiation operator ($D = d/dt$).

If a solution to the equation is sought in the exponential form

$$x = x_0 e^{zt} \tag{13}$$

then the characteristic equation is written as:

$$R(z) = a_m z^m + a_{m-1} z^{m-1} + \ldots + a_0 + (b_p z^p + b_{p-1} z^{p-1} + \ldots + b_0)e^{-z\tau} = 0. \tag{14}$$

The last equation does not contain a time variable and defines only a relationship between system parameters. The parameters can be found by analyzing data collected for a population at a certain time. In this connection, the characteristic equation (14) can be treated as a density function. Such functions are used for processing experimental data related to a certain time. We will show that basic density functions used in practice are governed by the characteristic equation (14) which also provides the development equation (5) as its particular case.

When finding a solution to Eq. (14) often used are estimations of weights laid on its terms. The general structure of a solution is determined by eliminating terms with lighter weights. For different ranges of variables and parameters, different terms appear to have a dominant significance. Therefore, when $z \gg 0$ it is assumed that

$$R(z) = a_m z^m + b_p z^p e^{-z\tau} = 0 \tag{15}$$

or, in the case where there is one dominant term among the functions of current time and one dominant term among those of delayed argument, we have

$$R(z) = a_{m-i} z^{m-i} + b_{p-j} z^{p-j} e^{-z\tau} . \tag{16}$$

In the above case the condition $z \gg 0$ may be insufficient, because emphasis is placed on the terms in the equation that have a dominant significance over a particular range of the system characteristics. For a small range, a development is governed by a specific mechanism and this manifests itself in the fact that a few terms in the equation of general type carry much greater weight than the rest. Variations in the set of dominant factors result in changes in the number of significant parameters. The detection of ranges over which the performance characteristics of a system do not vary enables us to make use of this sort of assumption. (The assumption is often made in practice for integrating differential equations.)

As the structure of characteristic equations (15) and (16) is the same, we may confine ourselves to the examination of only one of them.

When $m-p = \gamma > 0$, we get an equation of delayed type, when $p-m = \mu > 0$, the equation is of leading type. Characteristic equations are, respectively,

$$a_m + b_p z^{-\gamma} e^{-z\tau} = 0 \tag{17}$$

and

$$a_m + b_p z^{\mu} e^{-z\tau} = 0 .$$

Provided that $p = m$ (the differential equation is of neutral type), Eq. (15) is written as

$$z^m (a_m + b_p e^{-z\tau}) = 0$$

and the roots of the equation that are distant from the origin of coordinates can be roughly evaluated from the following equation

$$a_m + b_p e^{-z\tau} = 0 .$$

Note that basic distribution laws are derived from the characteristic equation (14) when $z \gg 0$ and when its parameters have particular values and its variables lay in a specific range associated with the expression (17) (see Table 1).

Table 1. Parameters of Eq. (17) and corresponding distribution functions

No	Equation parameters			Density function	Distribution law
	b_p	$p-m$	τ		
1	$\frac{1}{\sigma\sqrt{2\pi}}$	0	$\frac{z}{2\sigma^2}$	$\frac{1}{\sigma\sqrt{2\pi}} e^{-z^2/2\sigma^2}$	Gaussian
2	τ	0	τ	$\tau e^{-z\tau}$	Exponential
3	$\frac{1}{k!}$	k	1	$\frac{z^k}{k!} e^{-z}$	Poisson
4	$\frac{1}{C^k(k-1)!}$	$k-1$	$\frac{1}{C}$	$\frac{z^{k-1} e^{-z/C}}{(k-1)!\,C^k}$	Gamma
5	$\frac{k}{C}$	$k-1$	$\frac{z^{k-1}}{C}$	$\frac{k}{C} z^{k-1} e^{-z^k/C}$	Weibull
6	$\frac{1}{\sigma^2}$	1	$\frac{z}{2\sigma^2}$	$\frac{z}{2\sigma^2} e^{-z^2/2\sigma^2}$	Rayleigh
7	$\frac{1}{2^{k/2}\Gamma(k/2)}$	$\frac{k-2}{2}$	1/2	$\frac{z^{(k-2)/2}}{2^{k/2}} \frac{e^{-z/2}}{\Gamma(k/2)}$	Chi-square

Consider a differential equation which is associated with the characteristic equation (17)

$$x^{(m)}(t) = -\frac{b_p}{a_m} x^{(p)}(t-\tau) .$$

When allowance is made for (13), it becomes

$$z^m e^{zt} = -\frac{b_p}{a_m} z^p e^{zt} e^{-z\tau} , \tag{18}$$

Dividing both sides of Eq. (18) by z^{m-1}, we get

$$z e^{zt} = -\frac{b_p}{a_m} z^{p-m+1} e^{z(t-\tau)}$$

and again, taking account of (13), we find

$$\dot{x}(t) = -\frac{b_p}{a_m} z^{p-m+1} x(t-\tau) . \tag{19}$$

The equation is associated with the characteristic equation (15). Hence

$$\frac{b_p}{a_m} = -\frac{e^{z\tau}}{z^{p-m}}$$

and Eq. (19) is rewritten in the form

$$\dot{x}(t) = z e^{z\tau} x(t-\tau) .$$

Denoting $z e^{z\tau} = k(t)$, we again obtain the development equation (5).

Thus, density functions frequently used for processing experimental data are derived from the characteristic equation associated with the development equation (5), with its parameters having specific values. The parameters of the density functions can then be interpreted in terms of lag and growth rate of an associated process. It is known that a family of special functions is introduced in the fluctuation theory which serve as models of fluctuation processes (such as the Bessel functions, Legendre functions and so on that are governed by the hypergeometric equation) (Whittaker and Watson 1927)

$$(a_2 t^2 + a_1 t + a_0)\ddot{x} + (at + b)\dot{x} + cx = 0$$

with specified values of its parameters. Provided at $+b \neq 0$, we can write the hypergeometric equation in the form

$$(at+b)\left[\dot{x} + \frac{a_2 t^2 + a_1 t + a_0}{at+b}\ddot{x}\right] = -cx\ .$$

It is evident that the bracketed expression is two terms of the expansion of a function of delayed argument in a truncated Taylor series. As stated above, such an expansion is used for finding approximate solutions to delay differential equations (Elsgolts and Norkin 1971). Then the generalization of the hypergeometric equation is of the form

$$\dot{x}\left(t + \frac{a_2 t^2 + a_1 t + a_0}{at+b}\right) = \frac{c}{at+b}\, x(t).$$

Denote

$$\frac{a_2 t^2 + a_1 t + a_0}{at+b} = \tau(t)\ .$$

By shifting the argument on the both sides of the hypergeometric equation by $\tau(t)$ we again obtain the development equation.

Thus, the group of processes described by special functions is also included in the development equation as its particular case.

We now consider the connection between the development equation and the numerical measure of the amount of information introduced by Shennon (1963). As a measure of the amount of information generated by a Markov process, Shennon has introduced a concept of entropy for systems of events with probabilities $p_1, p_2, \ldots, p_n$

$$H = -\sum_{i=1}^{n} p_i \log p_i$$

where logarithms are taken to unspecified base. Such a measure of information has been derived owing to the fact that the following requirements are satisfied:

1) The function should be continuous in p_i.
2) When $p_1 = p_2 = \ldots = p_n = 1/n$ the function should increase monotonically with n. The uncertainty associated with equally probable events is greater than that for unequally probable events.

3) If a sample is broken up into two successive samples then the original entropy H should be a weighted sum of individual values of H.

The only function that satisfies the requirements is entropy. The function plays an important part in information theory, for it serves as a measure of the amount of information, as well as a measure of the occurrence of a sample and the uncertainty. It coincides with the expression for entropy in statistical physics. Here p_i is the probability that a system is in a location i of its phase space.

For a continuous distribution function $p(x)$, the entropy is defined by

$$dH/dx = -p \log p$$

and

$$H = -\int_{-\infty}^{\infty} p(x) \log p(x)\, dx \ .$$

Due to the fact that the unit of information is defined by the expression with the base of logarithms unspecified, we will further make use of natural logarithms. The rate of change of the entropy with respect to x is then given by

$$dH/dx = -p \ln p \ .$$

Consider the connection between the measure and the dynamic characteristics of a continuous process with entropy H. Write the above expression in the form

$$\exp\left(\frac{1}{p}\frac{dH}{dx}\right) = \frac{1}{p} \ .$$

Let K be the inverse of the characteristic time of a process with entropy H. Multiply both sides of the last expression by K.

$$\frac{K}{p} = K \exp\left(\frac{1}{p}\frac{dH}{dx}\right) \ .$$

Denote $K/p = z$. Then $1/p = z/K$ and

$$z = K \exp\left(\frac{1}{K}\frac{dH}{dx} z\right) \ .$$

Denote $\frac{1}{K}\frac{dH}{dx} = \tau$. Hence

$$z = K e^{z\tau} \ . \tag{20}$$

This is a characteristic equation associated with the linear differential equation

$$\dot{x} = -K x(t-\tau)$$

or

$$\dot{x} = K x(t+\tau) \ ,$$

corresponding to the development equation (5) or its particular version (4). Thus, entropy is defined by the characteristic equation associated with a particular version of the development equation. At the same time, this implies that the derivative of entropy with respect to x is proportional to the time shift of the argument, i.e., to the response lag or lead of a system.

Ashby (1962), when analyzing the adaptation mechanisms of ultrastable systems, emphasized that, when a system is faced with danger, the optimal duration of a test for changing its own parameters to be adapted is consistent with time required for the information to pass from step mechanisms, which impel the system to the test, to dominant parameters, which signalize test results. Thus, the nonlocal properties of a message play a major part in its ordering.

With its parameters having different values, the characteristic equation (20) governs basic distribution functions used in mathematical statistics, as well entropy which serves as a measure of the amount of information. This defines a fundamental connection between the entropy, as a quantity proportional to lag, and basic distribution laws. It is known that finding the extremum of the entropy function for limited moments (such as expectation, variance, etc.) results in extremals which are described by the distribution laws of mathematical statistics (Shennon 1963). Thus, the distribution laws of mathematical statistics can be achieved based on the variational principle and the properties of entropy defined by the characteristic equation for the development equation (5).

Entropy represents a measure of uncertainty. Uncertainty is associated with delay, i.e., time which elapses between the instant at which a system is exposed to an input action and that at which the system output begins to respond to it. For instance, after a management decision is made and implemented, its result becomes apparent only in the operation cycle time of a system. The uncertainty takes place in intervals. A measure of uncertainty is the delay of system response to an input signal.

The results obtained show that *the development equation (5) covers, as its particular cases, basic models of growth, mathematical statistics, fluctuation theory, and information theory. It also governs models used in the catastrophe theory.* This degree of generality became possible owing to the use of two axioms for formulating the equation: 1) development processes run as a branching process according to a chain mechanism and 2) for any development process, its prehistory is of great significance.

2.3 Hierarchy of Development Processes

As noted above, a development process begins with an exponential stage as a process of expanded reproduction which is defined by Eq. (4) or Eq. (1) with a constant growth rate. For a long period, a development proceeds with a decrease in its rate, yielding a model defined by Eq. (2). This model, called an allometric model or model of nonuniform growth, involves a decrease in growth rate with time and yields a straight line if the growth curve is plotted on a log-log sheet.

Introduce a new variable

$$T_1 = \ln t \ . \tag{21}$$

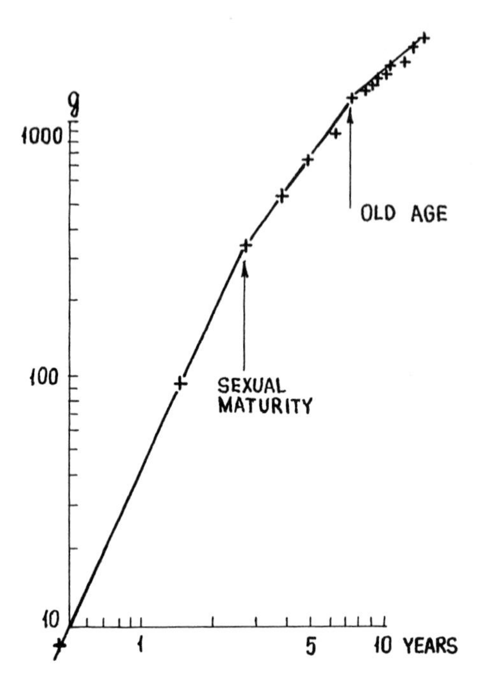

Fig. 6. Change in bream mass in the postembryonic development period. *Abscissa* age; *Ordinate* mass

The equation of allometric growth then assumes the form (1), i.e.,

$$\frac{dx}{dT_1} = k_1 x \ .$$

The logarithm of the complete age t of a system is laid out on the time axis. Logarithmic scales find application in biology to determine the age of a living system.

Development processes for which a change in system size occurs in line with the allometric model cover a substantially longer growth period, as compared to exponential intervals. But, as is the case with the exponential model, the scope of the allometric model is also limited. Figure 6 shows, on a log-log chart, how the weight of bream changes with its age. The curve includes three characteristic sections, the first of which terminates in maturity and the second in the coming of old age. The two moments in organism development are associated with changes in the regulatory and functional characteristics of the organism, in view of which they are called critical points.

An envelope of allometric modes is considered to form the next hierarchical level in system development, just as the allometric mode aggregates the sequence of exponential modes. The previous lower levels are defined by Eqs. (1) and (2). If the new hierarchical level is considered as one to be formed similarly to the lower levels, the process equation will be written as

$$\frac{dx}{dT_1} = \frac{k_2}{T_1} x \qquad (22)$$

or

$$\frac{d \ln x}{dT_2} = \frac{d \ln x}{d \ln T_1} = \frac{d \ln x}{d \ln \ln t} = k_2$$

where $T_2 = \ln T_1 = \ln \ln t$.

Hence

$$\frac{d \ln x}{d \ln t} = \frac{k_2}{\ln t}$$

and

$$\frac{d \ln x}{dt} = \frac{k_2}{t \ln t} . \tag{23}$$

A plot of system size on the $\ln x - \ln \ln t$ coordinates yields a straight line. In our example, model (23) presents changes in the mass (weight) of bream from birth to death (Fig. 7). Note that the stage of exponential growth is associated with processes occurring on the lower hierarchical levels. The allometric model, in the capacity of an envelope of sequential exponential stages, defines the development of an organism from its birth to its natural death.

Thus, a bank of models (1), (2), (23) is formed by introducing a sequence of relationships between growth rate and time as an independent variable. Model (1) then involves constant growth rates. In model (2), growth rates vary inversely with the independent variable, and in model (23), the growth rate is divided by the logarithm of an exponential function of time. (The same function defines entropy in statistical physics and information theory.)

To generalize, an envelope of the sequence of relationships presented by model (23) is written as

$$\frac{d \ln x}{dT_2} = \frac{k_3}{T_2} .$$

Then

$$\frac{d \ln x}{d \ln \ln t} = \frac{k_3}{\ln \ln t}$$

and

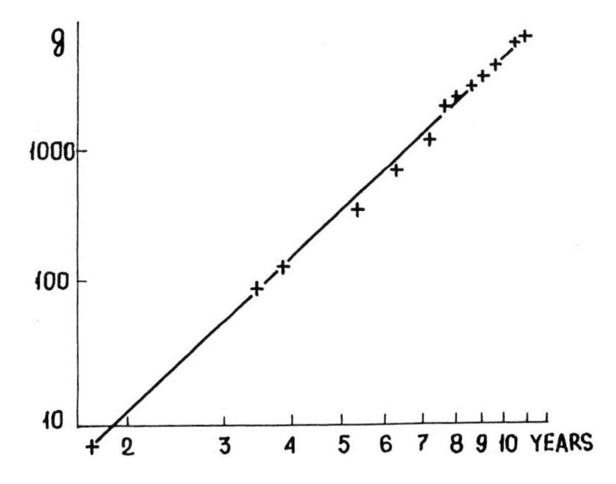

Fig. 7. Bream mass as in Fig. 6 but a lg-lg scale is chosen for *abscissa*

$$\frac{d \ln x}{d \ln \ln \ln t} = k_3$$

or

$$\frac{d \ln x}{dt} = \frac{k_3}{t(\ln t)(\ln \ln t)} .$$

For a system of the Nth level, we have

$$\frac{d \ln x}{dt} = \frac{k_N}{t(\ln t)(\ln \ln t)\ldots\underbrace{(\ln \ldots \ln t)}_{N-2}} .$$

As an example, we consider growth in the weight of chick embryo liver. Figure 8 shows data on growth dynamics plotted on the $\ln m - t$, $\ln m - \ln t$, and $\ln m - \ln \ln t$ coordinates.

It follows from the above that a transformation from an exponential mode to another takes place at the critical point. A sequence of exponential stages with decreasing growth rates is aggregated by the allometric relationship. Points of allometry inflection are instants of time at which the characteristics of a developing system change and these changes are more significant than those at the points of exponential bends. A sequence of allometric relationships in turn has its own envelope whose beginning and end define a complete development cycle on higher hierarchical level as against allometry.

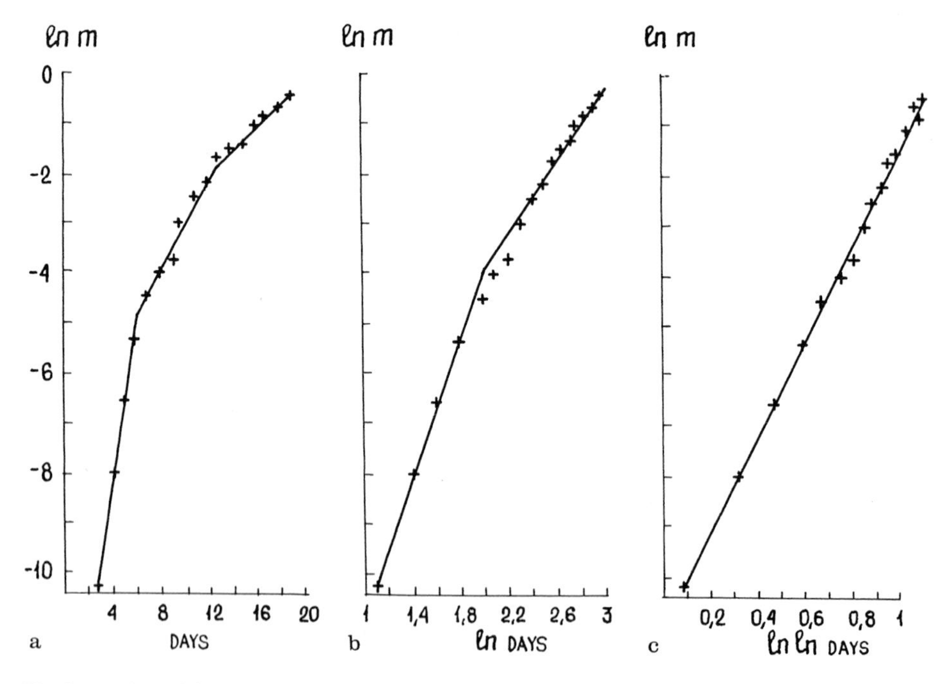

Fig. 8a–c. Growth in the mass of chick embryo liver. *Abscissa* time; *Ordinate* mass in mg (Data from Schmalhausen 1984, V. 1; p. 34)

Consider a scheme of preferable usage of time scales which are derived successively by taking the logarithm of a time scale on a preceding level. With this in view, we take advantage of a concept of system robustness. This property of a system suggests that a system trajectory experiences small variations with the variations of parameters on the right-hand side of differential equations which define the system or process (Andronov and Pontryagin 1937; Andronov et al. 1967; Bautin and Leontovich 1976).

We begin with considering a model of exponential growth $\dot{x} = kx$. The reproduction parameter k is assumed to vary in inverse proportion to age t raised to the power α.

$$\dot{x} = \frac{k_1}{t^\alpha} x \ .$$

When $\alpha = 0$, the last equation coincides with the exponential model. When α falls in the interval $0 \leqslant \alpha < 1$, it integrates to the following expression

$$\ln x = \frac{k_1}{1-\alpha} t^{1-\alpha} + \ln C \ .$$

Provided $\alpha < 1$, a collection of system trajectories $x = f(t)$ represents a set of ascending power functions of t. With $\alpha = 1$ the model turns allometric and is defined by a function of another class

$$\ln x = k_1 \ln t + \ln C \ .$$

When $\alpha > 1$, system trajectories $x = f(t)$ are descending functions of t. The value $\alpha = 1$ separates trajectories with different growth dynamics. Provided $0 \leqslant \alpha < 1$, information about critical points can be expected to be saved. When $\alpha = 1$, a new domain of robustness comes into being. The substitution of variable $\theta = \ln t$ turns an allometric model into an exponential one. As a consequence, all things are repeated and changes in the parameter α lead to the fact that the scale of the x-axis becomes the iterated logarithm of original variable, t, $\ln \theta = \ln \ln t$, and so on.

Thus, the life history of a system is divided into stages. Transitions between stages occur abruptly. This appears as an abrupt change in the rate of exponential growth or in the exponent of the allometric model. Points at which growth parameters change are referred to as critical points, because a morpho-functional reorganization and a change in the characteristics of a developing system take place at these points. The critical points are ordered with respect to their significance in the hierarchy. The significance of a critical point for the associated development process rises as the ratio of age and size intervals increases between successive critical points. i.e., as a hierarchical level has a higher priority in the sequence: exponent, allometry, an envelope of allometric models, and so on.

It is well known that an exponential dependence is widely used because it is associated with the rule of compound interest.

Processes of the allometric type are also discussed in detail in the literature. As noted above, Schmalhausen (1935) used a power function for modeling growth processes and called it a model of parabolic growth. He also ascertained changes in allometry parameters and attributed them to fundamental changes in organism development (see Fig. 1). Rosen (1969) pointed out that there are a variety of functionals defined on various sets of objects of both organic and inorganic nature. The functionals are defined by an equation of the following form

$$x = At^{B} .$$

The relationship which is represented by the equation and studied in detail by Huxley (1932) for living organisms, is called the law of heterogony or nonuniform allometric growth. This is the only type of explicit relationship between functionals concerning biological objects to which a detailed consideration was given.

By term „allometric development“ we mean a class of functionals defining relationships between any two characteristics of a process which is, in turn, defined by a power function, regardless of the specific meaning of its argument.

It is not possible here to use the term „parabolic growth“ coined by Schmalhausen, because we deal with a general case where the exponent of a power may be not only positive. Besides, the results obtained are concerned with general properties of processes whose model is defined by a power function. Therefore, the results can be of interest to scientists not only for examining the time history of a system, but also for solving problems in which an independent variable is not a time variable. For allometric relationships, a problem always arises which consists in determining which of two variables is an independent variable and which is a dependent variable.

In recent years, attention has again been paid to models of the allometric type which find application to processes of various nature (Von Bertalanffy 1952, 1968). Savageau's works (1979) are representative in this sense. One of the works, *Allometric Morphogenesis of Complex Systems: Derivation of Basic Equations from Primary Principles* came under the heading Organism and Population Biology, Social Systems, Economics and Management. The author notes with surprise what a large number of phenomena are defined by the simple equation, such as relationships between characteristics for main classes of animals and higher plants, data on morphology, physiology, pharmacology, biochemistry, cytology, evolution, and the ethology of some diseases.

For a long time allometry was considered to be an empirical law. But a good many works appeared in recent years, in which attempts are made to theoretically justify it. First of all the findings that are concerned with the theory of likelihood and dimensions. A power function is distinguished by its property that a ratio of two derived values is independent of the scale of basic units of measurement. In this connection, the function is used as a fundamental relationship in the theory of dimensions.

Studies in the theory of branching random processes resulted in establishing Zipf's law (1949), which is believed to be one of the main empirical laws of modern science. Consider a collection of N elements, with each element being marked with an index belonging to a certain set. When N is sufficiently large, each index appears exactly x times in a sample of N elements and the number $n(x, N)$ of different indices is

$$n(x, N) = A/x^{\gamma}$$

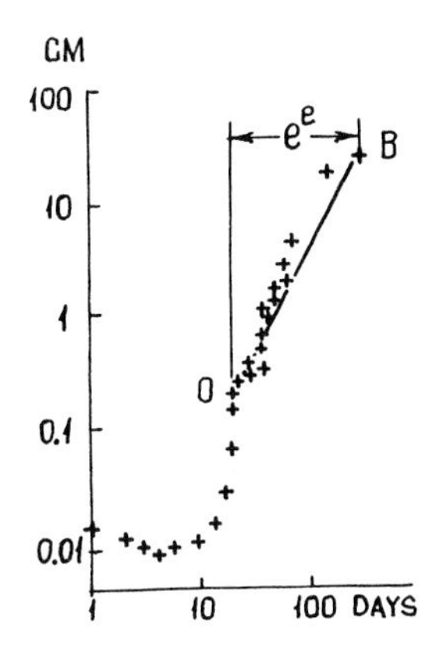

Fig. 9. Growth dynamics of human embryo in linear size; *0* beginning of organismic development; *B* birth; *Abscissa* embryo age reckoned from conception; *Ordinate* length. Logarithmic scales for both axes (Data from *Biology Data Book* 1964)

where A is the constant depending on sample size and γ the exponent of Zipf's law.

The law goes under the name of Zipf, Lotka, Mandelbrot or Pareto depending on its applications (Yablonsky 1977). Zipf's law is stationary like the law of normal distribution (a composition of stationary distributions is a distribution of the same type).

Rosen (1969) has derived an equation of allometric growth from the optimality principle by examining the class of functionals whose minimum solutions, such as the Eulerian equation, are power functions.

Savageau (1979) has proposed to develop the allometric model from a differential equation of the form

$$\dot{x}_1 = \alpha_1 x_1^{B_1} - \alpha_2 x_1^{B_2} ,$$

assuming x_1 to control the development of a system at a given allometric stage.

Generally speaking, growth rate may not only decrease, but also increase. Figure 9 shows data concerning linear growth of a man embryo (*Biology Data Book* 1964). As shown in the figure, at the gastrula stage (from 7th to 19th day), the growth proceeds at an increasing rate. Segments of this kind can also be observed on growth curves for other species. In particular, the growth of rat and pig embryos proceeds similarly at the gastrula stage (according to *Biology Data Book* 1964). An increase in growth rates is also noticed for processes of population growth (Bayley 1970; Watt 1971).

One can expect development at such stages to involve scale expansion, as distinct from the above scale-down mechanisms. To follow the above sequence of reasoning, growth curves with increasing rates are to be plotted on the $\ln \ln (x/C) - t$ coordinates for the initial development level, on the $\ln \ln \ln (x/C) - t$ coordinates for the next higher level, and so on. Then, for the initial level, the period of evolutionary development is defined by an equation of the form

$$\frac{dx}{dt} = k_1 e^{kt} x$$

or, after integration,

$$\ln x = \frac{k_1}{k} e^{kt} + \ln C \ .$$

Hence

$$\ln \frac{x}{C} = \frac{k_1}{k} e^{kt}$$

or

$$\ln \ln \frac{x}{C} = kt + \ln \frac{k_1}{k} \ .$$

It can be said that, unlike the above case of decreasing growth rates where a time scale contracts in going from a hierarchical level to another higher level, such contraction takes place but in the y direction. In contrast to a linear scale, the time scale then expands.

For the origin of the process to match to ordinate zero, data are to be plotted on the $\ln \ln (x/x_0) - t$ coordinates, ensuring $\ln \ln e = 0$ at $x = x_0$. Data on the length of man embryo (Fig. 9) replotted on these coordinates are shown in Fig. 10. The growth from the 9th to the 19th day is actually seen to go on according to the model under consideration.

A typical example of growth with increasing growth rates is the dynamics of world population from the middle of the seventeenth century up to the present (Fig. 11). As follows from Fig. 12, the tendency changed since the 1950–60's. For 300 years up to this time the growth of population in Europe and Asia went on according to relationships which were close both to each other and to the dynamics of world population. Since the 1950–60's the situation has been changing – a sharp tendency to stabilization of the population in European countries becomes apparent against the background of an increase in world population.

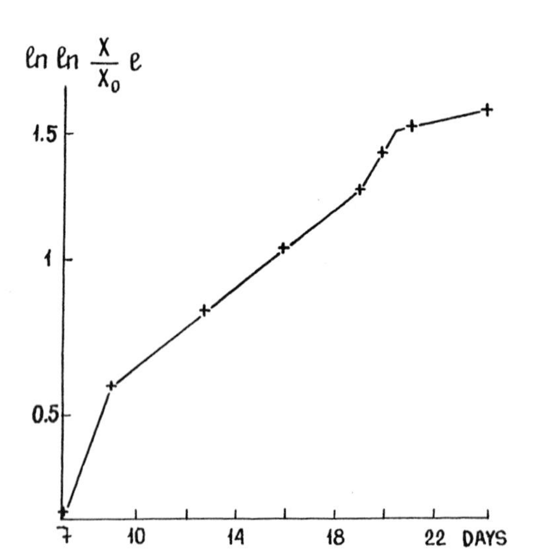

Fig. 10. Growth dynamics of the length of the human embryo at early development stages. *Abscissa* age; *Ordinate* $\ln \ln (x/x_0)$ e (Data from *Biology Data Book* 1964)

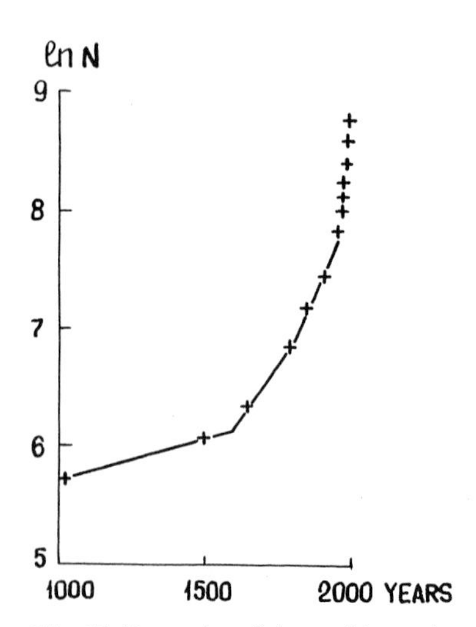

Fig. 11. Dynamics of the world population. *Abscissa* time; *Ordinate* the logarithm of population size (Data from Urlanis 1978)

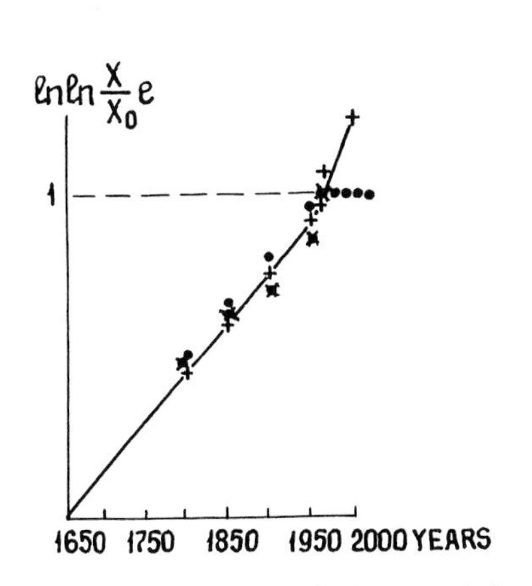

Fig. 12. Dynamics of the population: ● Europe; × Asia; + world population. *Abscissa* time; *Ordinate* $\ln \ln (x/x_0)\, e$ (Data from Urlanis 1978)

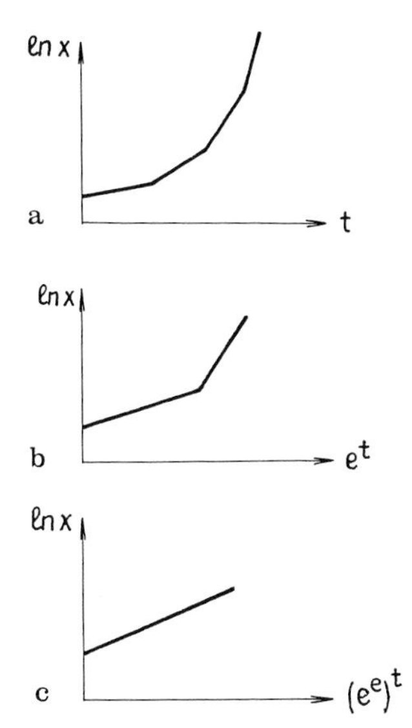

Fig. 13a–c. A sequence of scales for processes with increasing growth rates

The change of tendency in development occurs when the ordinate assumes a value close to unity.

We will use negative numbers to denote the hierarchical levels on which a process progresses at the increasing growth rate, assuming a tendency in development to change when it passes through the zero level. We then get a sequence of coordinates for different hierarchical levels, as shown in Fig. 13a–c.

Note that such a type of relationship as shown in Fig. 9 is met with in the reliability theory, experimental design, and growth kinetics of biological systems. For example, the reliability theory uses an iterated exponential law with the following distribution function

$$F(x) = e^{-e^{-x}} .$$

The distribution law is used in redundancy with similar problems, in particular for the scheme of cumulative damages to many components (Kordonsky 1963). In the theory of experimental design, a transform function (Adler et al. 1976)

$$d = 1 - e^{-e^{-x}}$$

is introduced, which is regarded as preferable owing to its higher sensitivity in the middle region, as compared to that in the regions close to 0 and 1.

Thus, many scientists show an interest in the use of e^e as a particular logarithmic base.

Taken together, the above scales permit the structuring of information about the development of natural systems, with evolution periods and critical points for each structure coming to light.

Thus, the complete set of axioms permits building a model of development of natural systems based on the facts that development is defined as a branching process, a development course is dependent on its prehistory, and the coupling between hierarchical levels is determined by taking the natural and inverse logarithms of independent variables. The set of axioms allows one to quantify critical constants of such systems. Comparing the obtained results with actual data on the development of natural systems represents a peculiarly fascinating exercise, which is discussed in another paper (Zhirmunsky and Kuzmin 1986).

We will consider below how critical constants are estimated for each of the hierarchical levels and how they interact with each other.

3 Model of Critical Level Constants in System Development

This section deals with critical constants based on the analyses of the development equation. The critical constants comprise a hierarchy of critical constants and define time intervals in which a system retains its qualities on different hierarchical levels. The basis of the hierarchy is formed with power-exponential functions of Napierian number e

$$\ldots (1/e)^{(1/e)},\ 1/e,\ 0,\ 1,\ e,\ e^{e},\ e^{(e^{e})},\ldots$$

The requirement to synchronize the critical boundaries defines a development unit. The examination of many facts concerning the development of natural systems shows that the hierarchy of critical constants and their synchronization permit it to reveal significant information about main mechanisms of forming time and space rhythms, as well about the structures of natural systems belonging to different hierarchical levels.

3.1 Some Peculiarities of the Development Equation

A main problem to be tackled is whether reorganization phases and points which can be brought out from morpho-functional features or from changes in the phase state, can be determined from the development equation. Can one find the moment at which a system undergoes qualitative changes, given a model of its development? Moreover, how are critical points to be located if the location of one critical point is known?

With this end in view, we will consider the properties of solutions to the particular case (4) of the development equation and then proceed to its extension

$$\dot{x} = k\,x\,(t-\tau) \tag{24}$$

with $k>0$, $\tau>0$.

Let a solution of the equation be sought in the form (13) where $z=u+iv$. Hence, a characteristic equation gets the form

$$z e^{zt} = k\,e^{z(t-\tau)}$$

or

$$z \quad = k\,e^{-z\tau}\ . \tag{25}$$

From the Euler formula

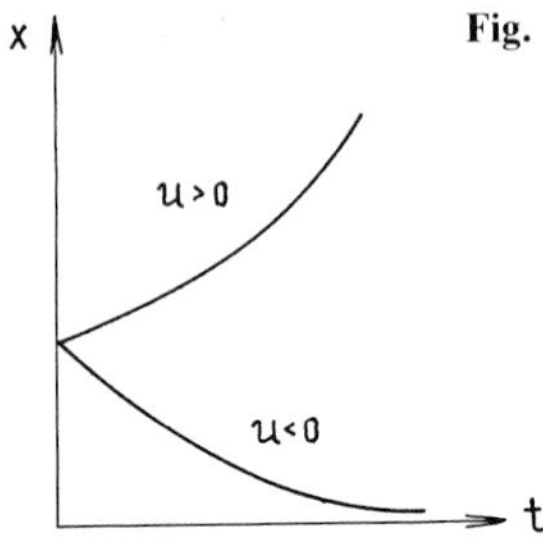

Fig. 14. Solution of Eq. (4) when $v = 0$

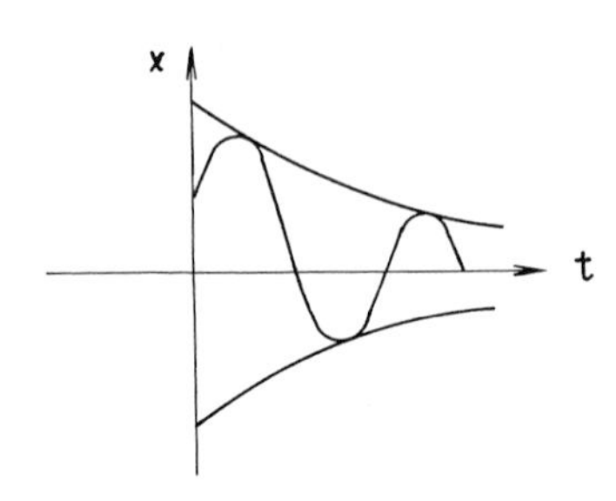

Fig. 15. Solution in Eq. (4) provided $u<0$, $v = 0$

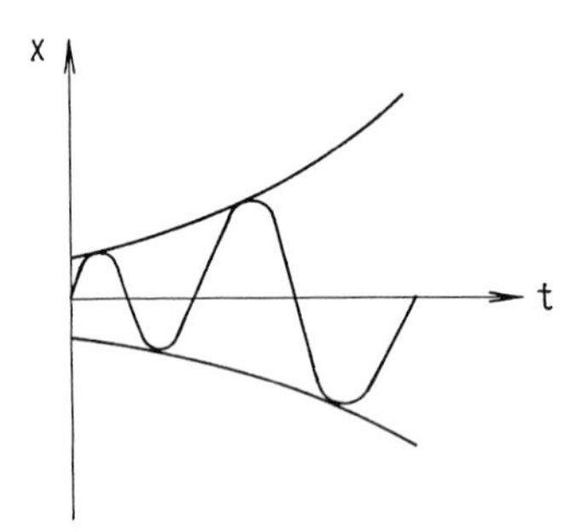

Fig. 16. Solution of Eq. (4) when its roots are pseudo-positive ($u>0$, $v \neq 0$): a prohibited mode

$$e^{-iv\tau} = \cos v\tau - i \sin v\tau \ ,$$

we have

$$u + iv = k e^{-u\tau}(\cos v\tau - i \sin v\tau) \ .$$

By splitting the equation into the real and imaginary part, we get

$$u = k e^{-u\tau} \cos v\tau \ , \tag{26}$$

$$v = -k e^{-u\tau} \sin v\tau \ . \tag{27}$$

For nonoscillating processes (aperiodic mode), $v = 0$ and the growth curve has the form as shown in Fig. 14. When $v \neq 0$ and $u<0$ we have a damped oscillatory mode in which the amplitude of the oscillating quantity decreases with time (Fig. 15). When $v \neq 0$ and $u>0$ (such roots are called pseudopositive roots), oscillations have an exponentially increasing amplitude (Fig. 16).

For a nonoscillating process we get from (26)

$$u = k e^{-u\tau}$$

or, on multiplying the left and right sides by τ,

$$u\tau = k\tau e^{-u\tau} \ . \tag{28}$$

The graphical solution of the equation is shown in Fig. 17. Here the solution is unique and dependent on the value of k. Hence, there exists a single exponential growth mode ($u>0$) when $v = 0$

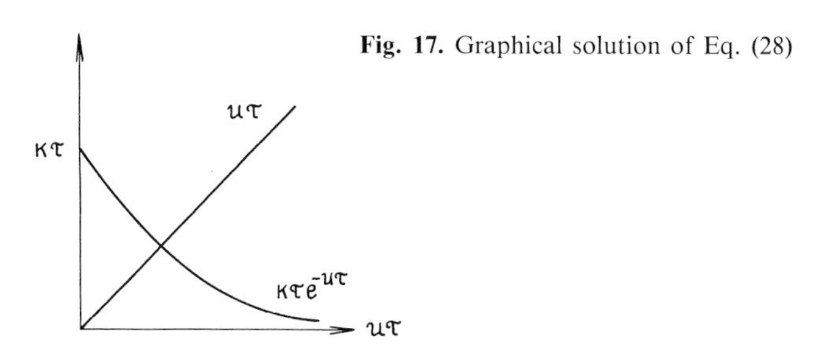

Fig. 17. Graphical solution of Eq. (28)

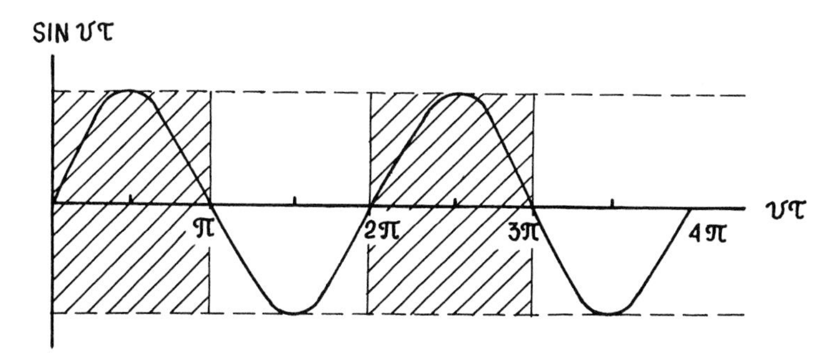

Fig. 18. The unequality $u>0$ $(v \neq 0)$ is not valid in the *dashed areas*

$$x = x_0 e^{ut} .$$

If $v \neq 0$ and $u>0$, the system undergoes oscillations with exponentially increasing amplitude, which is able to destroy it. It means that, to ensure a stable growth, the system must develop in the region where exponential growth takes place but undamped oscillations are impossible. We will find a region in the parametric space of Eq. (24), in which the requirement is satisfied.

From (27) we have

$$e^{u\tau} = -k\tau \frac{\sin v\tau}{v\tau}$$

or, after taking the natural logarithms of both sides,

$$u\tau = \ln k\tau + \ln\left(-\frac{\sin v\tau}{v\tau}\right) . \tag{29}$$

As the logarithm of a negative number does not exist, the inequality $u>0$ is valid only if the value of $v\tau$ ranges between π and 2π, 3π and 4π, and so on (Fig. 18).

Thus, at least in the interval $0<v\tau<\pi$, there is no root of the characteristic equation which would have a positive real part and a nonzero imaginary part.

Dividing (26) by (27), we get $u/v = \operatorname{ctg} v\tau$ or $u\tau = v\tau \operatorname{ctg} v\tau$. Therefore, for the real part of the root to be positive the condition

$$-v\tau \operatorname{ctg} v\tau > 0$$

must be fulfilled. Then $v\tau \neq 0$ and $\operatorname{ctg} v\tau < 0$. The limiting case where $\operatorname{ctg} v\tau = 0$ is associated with the appearance of a root with positive real part. Then the condition of existence of pseudopositive roots yields

$$v\tau = \tfrac{3}{2}\pi + 2\pi n, \quad n = 0, 1, 2, \dots \tag{30}$$

According to (29), $u\tau \geqslant 0$ is equivalent to

$$k\tau \geqslant -\frac{v\tau}{\sin v\tau}$$

or, considering (30), we get for the limiting case

$$k\tau = v\tau = \tfrac{3}{2}\pi + 2\pi n, \quad n = 0, 1, 2, \dots .$$

Thus, $k\tau = \frac{3}{2}\pi$ is the first value, beginning at which undamped oscillations occur in the system. If we substitute this value into Eq. (28), we get

$$u\tau = \tfrac{3}{2}\pi e^{-u\tau} .$$

The graphical or numerical solution of the latter yields

$$u\tau = 1.293 \tag{31}$$

when

$$k\tau = \tfrac{3}{2}\pi . \tag{32}$$

This illustrates that the parameter space of Eq. (24) has a boundary (31) below which purely exponential growth takes place and above which undamped oscillations occur in the system defined by (24) (Fig. 19). While the parameters of a developing system are within the region I, a stable exponential growth is possible. Passing through the boundary (31) into the region II gives rise to undamped oscillations. As a result, growth is impossible in the region II.

If the value of delay changes slowly in the course of growth, the exponential growth will take place below the boundary (31). The scheme in Fig. 20 illustrates how the system reaches the boundary by the time t_1 if the value of delay changes with time. It means that the exponential growth may continue up to moment t_1.

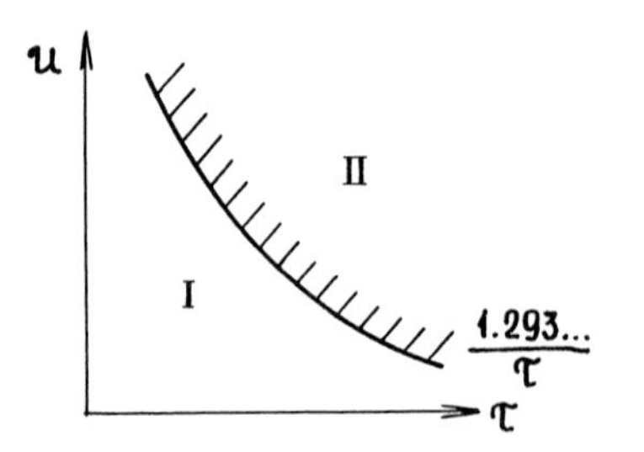

Fig. 19. *I* area of exponential growth; *II* area of unstability

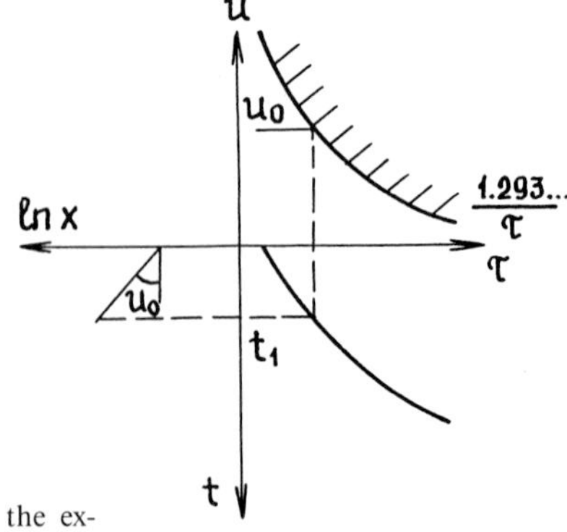

Fig. 20. Scheme for determining an instant at which the exponential growth terminates with a change in time delay

To stabilize a system when it reaches the boundary (31) it is necessary to decrease either growth rate or delay, or both growth rate and delay, to make the system come back into the region of stable growth.

An increase in oscillation amplitude, when the system gets to the forbidden region, manifests itself by a wider spread in experimental data. This is a general property of critical points for processes of a different nature. Thus, for example, Patashinsky and Pokrovsky (1975) note that, as a critical point is being approached, fluctuations occurring in a substance increase and the fluctuation phenomena bear a certain resemblance in various critical points. Uryson (1973) notes that individual differences in dimensions of a man body gain in the postembryonic growth period when a growth rate increases. In this connection, "the increased variability of morphological features at the ages related to the sexual maturation period may be attributed to certain biological factors, in particular to the effects of differences in physiological age when passing through critical points" (p. 26). The same point of view was taken by Thompson (1945).

At critical points, an increased susceptibility of a system to environmental factors is observed (Patashinsky and Pokrovsky 1975) and a correlation arises between system components, which does not disappear on a macroscopic scale.

The boundary (31) isolates the region of the parameter space in which a development process progresses in a sufficiently stable fashion. At the same time, a development tends to proceed in the region very close to the boundary (31) because this ensures maximum relative growth; as a result, the system can increase to a larger size. Mechanisms of selecting individuals by maximum relative growth have been noted in experimental data (Routh and Routh 1964): a hormone has been extracted from tadpoles of the leopard frog (*Rana pipiens*), which inhibits slowly developing individuals, resulting in their death.

The question arises, to what extent can the boundary be diverged from, so that a system could continue to develop? The answer is that the system needs to be reorganized so that, in the future, it can grow steadily for some time. This requires the actuating of mechanisms for adapting the system to future environmental conditions, since any control action is based on prediction. Thus, the reorganization mechanism to ensure a stable development of the system in the future must be a leading mechanism, requiring the growth rate of the system to be proportional to its future size. We write a differential equation for this mechanism

$$\dot{x} = k x(t+\tau) . \tag{33}$$

To solve the equation, an initial function $x(t) = \varphi(t)$, $t \in [t_0, t_0+\tau]$, must be known.

Anokhin (1962) believed that the leading reflection of the objective reality is a property of living matter which allows biological systems to get adapted to the medium. Such a property is called preadaptation (Henderson 1979). As a rule, a future state is predicted by extrapolating an established tendency to the future, with the past things taken into account. But abrupt changes in development are inconsistent with this course of reasoning. The longer the period over which a single tendency is observed, the shorter the rest of the period over which the established tendency will remain. The coexistence of leading and delay mechanisms is defined by the Hamiltonian equations, according to which the

Hamiltonian function for a process with dynamics defined by Eq. (24) has the form

$$H = \psi(t)\,k\,x(t-\tau)$$

and the equation for the conjugate variable is written as

$$\dot{\psi}(t) = -\frac{\partial H}{\partial x} = -k\,\psi(t+\tau) \ .$$

Thus, for the system whose development is defined by the delay differential equation, its conjugate variable is defined by the leading differential equation. This illustrates that, together with delay mechanisms, there exist mechanisms of leading type in system development.

Consider the structure of a solution to Eq. (33). Its characteristic equation is written as

$$z e^{zt} = k\,e^{z(t+\tau)} \quad \text{or} \quad z\tau = k\tau e^{z\tau} \ . \tag{34}$$

When $v = 0$, $u > 0$, as stated above, an exponential growth mode takes place. Figure 21 shows variants of the graphical solution of the equation. It is apparent from Fig. 21 that the maximum value of $k\tau$ at which a solution of the characteristic equation (34) is existent is equal to

$$k\tau = 1/e \tag{35}$$

at

$$u\tau = 1 \ . \tag{36}$$

The exponential growth modes for an equation of the leading type are shown to be sustained in the region below the boundary (36).

From (31) and (36), we obtain a region in which modes of stable growth may take place (Fig. 22). To stabilize the characteristics of a process when it reaches to boundary (31), process parameters must be changed so that the system will be in the region where the parameters of delay and leading mechanisms are matched. It means that, to achieve the stabilization of the system, it is necessary to carry out an abrupt change in the relative growth (by a factor of no less than 1.29), i.e., the jump between the boundaries of the region of stable growth yields the following rate ratio

$$u_1/u_2 = 1.29 \ .$$

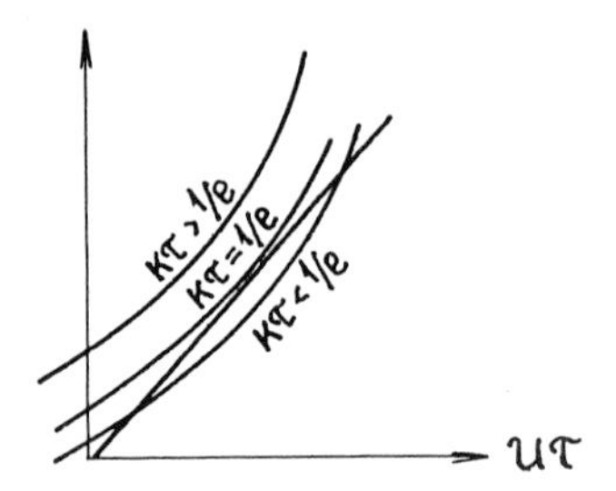

Fig. 21. Graphical solution of Eq. (34)

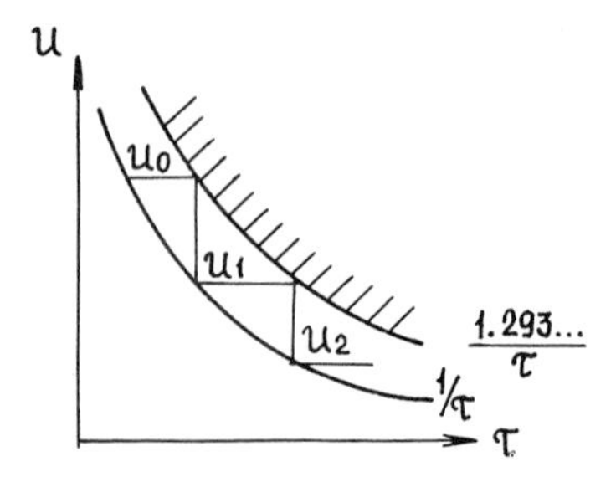

Fig. 22. Scheme of forming the rates of stable growth

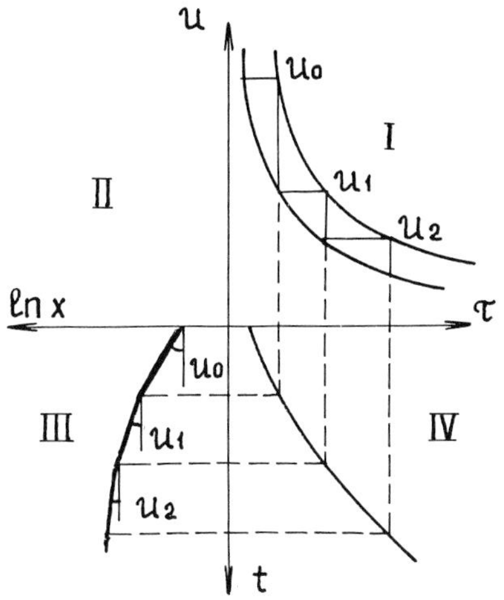

Fig. 23. Scheme of constructing a growth curve

This allows the character of system growth and the locations of critical points to be determined if an initial growth rate, the relationship $\tau = f(t)$, and the initial system size are known (Fig. 23). The growth rate is derived from the growth curve plotted on a semilog chart as the slope of the straight-line segment of the curve with the abscissa axis. On measuring the slope one can plot its value in the first quadrant (Fig. 23). The value of the exponential growth rate can remain unchanged only to the boundary (31), which specifies the value of delay when the growth will change. From $\tau = f(t)$ in the fourth quadrant (Fig. 23), the value of delay determines the occurrence of a critical point at which an exponential growth mode with given initial growth rate ceases. On decreasing the initial growth rate by a factor of 1.29, we obtain a new value for it which determines a slope in the third quadrant (Fig. 23).

Thus, the dependence of delay value on internal and external factors determines growth trends and the occurrence of critical phases in the exponential development. With the structure of the dependence found out, one can use it to analyze and predict the characteristic features of development processes (Kuzmin and Lenskaya 1974; Chuev et al. 1975; Kobrinsky and Kuzmin 1981).

The development equation (24) is of unstable type. Development processes contain, together with stages of growth, stages at which system size decreases. They are defined by a development equation of stable type

$$\dot{x} = -kx(t-\tau)$$

where $k > 0$, $\tau > 0$. We will seek a solution of the equation in the exponential form. Then a characteristic equation has the form

$$z = -ke^{-u\tau}.$$

In the case where oscillations are impossible ($v = 0$), the equation becomes

$$u = -ke^{-u\tau} \tag{37}$$

and its graphical solution is shown in Fig. 24. The maximum value of k at which a solution is existent is as follows

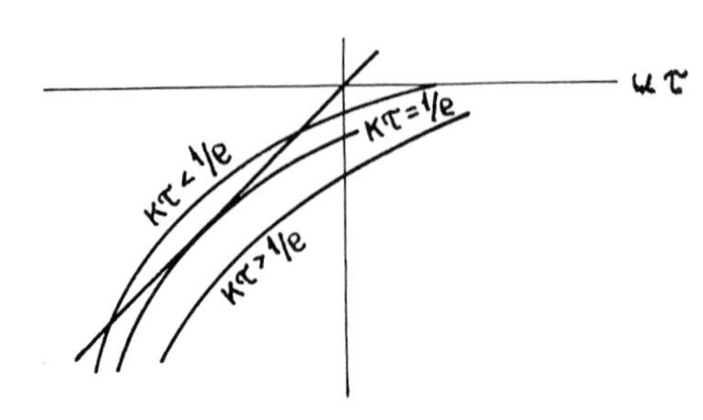

Fig. 24. Graphical solution of Eq. (37)

$$k\tau = 1/e \ . \tag{38}$$

Then

$$u\tau = -1 \ . \tag{39}$$

Thus, relationships (31) and (32) impose constraints on the parameter space of the development equation of unstable type, and relationships (38) and (39) on the parameter space of the development equation of stable type.

Relationships (31) and (39) impose restrictions on the applicability of distribution functions used in mathematical statistics. Therefore, the relationships are used for processing experimental data to find regions in which statistical populations are uniform (Evtikhiev and Kuzmin 1982; Kobrinsky and Kuzmin 1981).

3.2 Critical Levels of Allometric Development

This section discusses the general properties of the development model of stable type, which is defined by an equation of the form (5)

$$\dot{x} = -k(t)\,x[t-\tau(t)] \ . \tag{40}$$

A solution of Eq. (40) will be sought in the form

$$x = x_0 e^{-z(t)t} \ . \tag{41}$$

Because $k(t)$ and $\tau(t)$ are time-dependent, the rate of change of system size is also sought as a function of time.

From Eq. (41)

$$\dot{x} = -x_0[\dot{z}(t)t + z(t)]e^{-z(t)t} \ , \tag{42}$$

and the substitution of (41) and (42) into (40) yields

$$\dot{z}(t)t + z(t) = k(t)e^{-(t-\tau)z(t-\tau)+tz(t)} \ . \tag{43}$$

Consider the case where

$$t_* = \tau \tag{44}$$

and

$$\dot{z}(t_*) \leqslant 0 \ . \tag{45}$$

Expression (44) is generally regarded as a critical condition. The value of delay cannot be greater than the current age of the system. Inequality (45) defines the class of processes whose growth rates decrease with system age.

From (43) and (44)

$$\dot{z}(t_*)t_* + z(t_*) = k(t_*)e^{z(t_*)t_*} \ . \tag{46}$$

Considering (45), Eq. (46) is rewritten as

$$\dot{z}(t_*)t_* = -z(t_*) + k(t_*)e^{z(t_*)t_*} \leqslant 0 \ . \tag{47}$$

From (47)

$$z(t_*) \geqslant k(t_*)e^{z(t_*)t_*} \ . \tag{48}$$

By multiplying both sides of (48) by t_*, we get

$$z(t_*)t_* \geqslant k(t_*)t_* e^{z(t_*)t_*} \ .$$

The graphical solution (see Fig. 21) shows that

$$z(t_*)t_* = k(t_*)t_* e^{z(t_*)t_*} \ .$$

Then

$$k(t_*)t_* = 1/e \tag{49}$$

and

$$z(t_*)t_* = 1 \ . \tag{50}$$

Hence, by substituting (50) into (41), we get

$$x_*/x_0 = 1/e \ . \tag{51}$$

From (49) und (50)

$$z(t_*)/k(t_*) = e \text{ or } z(t_*) = k(t_*)e \ . \tag{52}$$

Relationships (49) and (50) define limiting development conditions, since the product of time-dependent parameters $k(t)$ and $\tau(t)$ cannot be greater than the value specified by (49). Therefore, when the value of $k(t)$ reaches the boundary (49), the mode of stable growth ceases.

We multiply the left and right sides of (52) by t_* and substitute the result into (41). As (52) gives a lower bound, we get

$$\frac{x_*}{x_0} \geqslant e^{-z(t_*)t_*} = e^{-ket_*} \ . \tag{53}$$

As mentioned before, the allometric processes are frequently found. Then $k(t) = B/t$, i.e., the relative growth varies in inverse proportion to the current age of the system. Substituting $k(t) = B/t$ into inequality (53), we get

$$x_*/x_0 \geqslant e^{-Be} \ . \tag{54}$$

As a result, the lower bound to the stable growth curve proves to be a constant which is dependent on the parameter B of the associated allometric process, i.e., on the slope of the linear section of the growth curve plotted on a semilog chart.

The upper bound is obtained by examining the case where there is no delay. The process is then defined by the equation

$$\dot{x} = -\frac{B}{t}x$$

whose solution is the following power function

$$x = At^{-B} .$$

Thus, the upper bound is expressed as

$$\frac{x_*}{x_0} \leqslant \left(\frac{t_*}{t_0}\right)^{-B} . \tag{55}$$

From (54) and (55)

$$e^{-Be} \leqslant x_*/x_0 \leqslant (t_*/t_0)^{-B} .$$

Then

$$t_*/t_0 \leqslant e^e = 15.15426\ldots . \tag{56}$$

Consequently, the ratio of the argument values related to the beginning and end of stable allometric growth is constant and equal to e^e (Kuzmin and Zhirmunsky 1980a).

Studies of Eq. (40) in terms of its stability with the help of Krasovsky's theorem (1959) show that the stability of simple and asymptotic solutions is provided up to the critical point (Zhirmunsky and Kuzmin 1982). For the allometric mode, we have from (49)

$$B = 1/e . \tag{57}$$

Relationship (57) shows that for the basic variable the exponent B (the slope of the growth curve on logarithmic coordinates) is invariable and equal to 1/e. Thus, this sort of relationships determine the values of the function and its argument at the sequential critical points and the system behavior between them. This is closely related to the theory of critical phenomena (Fisher 1968; Patashinsky and Pokrovsky 1975; Ma 1980). A hypothesis of similarity is accepted in the theory of critical phenomena (Patashinsky and Pokrovsky 1975). According to the hypothesis, one parameter, namely a characteristic length, is chosen out of a great number of characteristics on which the critical phenomena have an effect. The dependence of the parameter on a temperature difference, $T - T_c$, is investigated. Here T and T_c are the current and the critical temperature, respectively. The expressions with describe the behavior of the rest of the variables or parameters at the critical points are deduced from the relationships of dimensional theory. At present, this approach is of great interest since attempts have been made at using a single standard, namely the standard of time, to establish a unified system of measurements, the rest of the physical units to be derived by using the relationships like power functions based on the fundamental constants.

As known, the physical constants relate different states of a system, with the states belonging to the same or different hierarchical levels. Hence, the values of the exponent of a power can be expected to demonstrate which level in represented by data analyzed. For example, the relationship between the weights of an animal's body and of its skeleton is defined by the equation in which $B = 1$. It means that the two parameters belong to the same hierarchical level. For the relationship between the lengths of a man's body and of his head, $B = 2.6$, i.e., the parameters belong to different hierarchical levels (the body length is the parameter of the whole, and the head length is the parameter of its constituent).

According to the results of experimental research, which were reported by Fisher (1968) and Ma (1980), the indices in the power dependences of heat capacity and spontaneous magnetization on $T-T_c$ fall within the ranges 0.11 – 0.17, 0.13 – 0.19, 0.15 ± 0.015, 0.14 ± 0.06, 0.07 – 0.14, for the first case, and within the ranges 0.32 – 0.36, 0.32 – 0.39, 0.312 ± 0.03, 0.37 ± 0.04, for the second case. It would be noted that the values are close to $1/e^2 = 0.135$ and $1/e = 0.367$, respectively. We can be convinced that the rest of the parameters on which the experimental data are available are consistent with the critical relationships that are discussed in full detail in Sect. 3.4.

Thus, the value of the exponent of a power conveys information to what critical level a parameter in question belongs. Therefore, two parameters of the same structural level do not convey new information. Thus, the satisfaction of condition (57) for the allometry level is an indication of the basic variable.

The solution of the allometric equation of unstable type

$$\dot{x}(t) = k(t)\,x[t-\tau(t)] \tag{58}$$

will be sought in the form (13)

$$x = x_0 e^{z(t)t} .$$

The characteristic equation is then written as

$$\dot{z}(t)t + z(t) = k(t)\,e^{(t-\tau)z(t-\tau)-z(t)t} .$$

Inequalities (44) and (45) will be assumed to hold. Using inequality (44), we get

$$\dot{z}(t_*)t_* + z(t_*) \geqslant k(t_*)\,e^{-z(t_*)t_*} . \tag{59}$$

Considering (45), inequality (59) is rewritten as

$$0 \geqslant \dot{z}(t_*)t_* \geqslant -z(t_*) + k(t_*)\,e^{-z(t_*)t_*}$$

from which

$$z(t_*) \geqslant k(t_*)\,e^{-z(t_*)t_*} . \tag{60}$$

As mentioned before, expression (60) in the capacity of equality has pseudo-positive roots when $k = 3/2\,\pi$ [refer to Eq. (32)]. It is clear that inequality (60) holds for this value of k. Then $z\tau = 1.293$ [refer to Eq. (31)] and

$$\frac{z}{k} = \frac{2 \cdot 1.293}{3\,\pi}$$

or

$$z = \frac{2 \cdot 1.293}{3\,\pi}\,k . \tag{61}$$

The relationship is correct at the critical point which is followed by the mode of oscillations with exponentially increasing amplitude. The argument value at the point is again denoted with t_*. We multiply the left and right sides of (61) by t_* and substitute the result into (13). Then

$$\frac{x_*}{x_0} \geqslant e^{[(2\cdot 1.293)/(3\,\pi)]\,k\,t_*} .$$

For the allometric mode, $k(t) = B/t$. So

$$\frac{x_*}{x_0} \geqslant e^{[(2\cdot 1.293)/(3\,\pi)]\,B} . \tag{62}$$

On the other hand, the upper bound to the solution of the equation of unstable type is obtained from the equation of allometric growth with out delay:

$$\dot{x} = \frac{B}{t}x$$

from which

$$x = A t^B;$$

the curve will not trace above the curve of pure allometric growth:

$$x_*/x_0 \leqslant (t_*/t_0)^B . \tag{63}$$

From (62) and (63)

$$e^{[(2\cdot 1.293)/(3\,\pi)]\,B} \leqslant x_*/x_0 \leqslant (t_*/t_0)^B$$

or, after taking the Bth root,

$$\frac{t_*}{t_0} \geqslant e^{(2\cdot 1.293)/(3\,\pi)} .$$

As τ decreases, the ratio of ages at two sequential critical points tends asymptotically to

$$\frac{t_*}{t_0} = e^{\,(2\cdot 1.293)/(3\,\pi)} = 1.316\ldots . \tag{64}$$

Thus, with delay small, two sequential allometric modes of stable and unstable types cover the age range which is defined by the product of (56) and (64)

$$t_*/t_0 = 15.154 \times 1.316 = 19.943 \tag{65}$$

(Fig. 25).

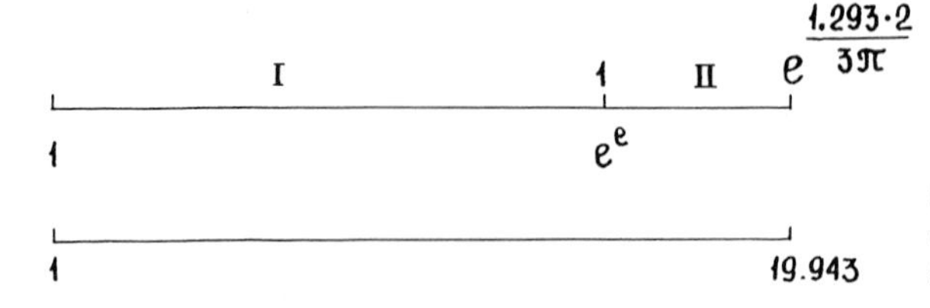

Fig. 25. Scheme of forming the sequential allometric processes of stable (*I*) and unstable (*II*) types

3.3 Critical Levels in Exponential Growth Models

Under invariable internal and external growth conditions, the development is known to proceed according to the exponential law. As shown in Sect. 3.1, with the internal and external conditions varying within some limited range (depending, in a sense, on the characteristics of the developing system itself), the development also remains exponential if certain restrictions on the parameters of the process or system are not violated.

Introduce a dimensionless variable

$$t/t_0 = \theta \; . \tag{66}$$

The power dependence is then rewritten as

$$x = A t_0^B \theta^B = A_1 \theta^B \; .$$

Introduce a new variable

$$\eta = \ln \theta \; , \tag{67}$$

which is also dimensionless. Hence,

$$x = A_1 e^{B\eta} \; .$$

This is an exponential dependence as well.

Because for the allometric process of stable type the ratio of ages at two sequential critical points is $\theta_* \leqslant e^e$, we get

$$\eta_* = \ln \theta_* \leqslant e \; , \tag{68}$$

i.e., for the exponential function the ratio of argument values at two sequential points does not exceed the Napierian number e. We will make several observations concerning the verification of relationship (68).

Consider the development equation (40) of stable type at the critical point where delay is equal to the system's current age

$$t_* = \tau \; .$$

The characteristic equation (43) for the critical point is then rewritten as

$$\dot{z}(t_*) t_* + z(t_*) = k(t_*) e^{z(t_*) t_*} \; .$$

The critical mode condition (48) then becomes

$$z(t_*) t_* = 1$$

and the substitution of this into (41) leads to

$$x_*/x_0 = 1/e \; . \tag{69}$$

It means that at the critical point where $t_* = \tau$ and the critical mode condition (50) is met, the ratio of system sizes at two sequential critical points is equal to the Napierian number e.

Notice that the relationships (31) and (50) are consistent with the empirical rule that is widely applied in the dimensional theory. By the rule, the value of a dimensionless variable at the vicinity of a certain critical point is of the order of unity (Migdal 1975).

It is known (Ashbi 1962; Guretsky 1974) that the development of a delayed system involves a regulation mechanism (the adaptation for new development conditions) having an operation period which is equal to time delay. This relates to the method for estimating critical states for the case where the value of delay at the critical point is close to the system's current age (Patashinsky and Pokrovsky 1975).

Relaxation dynamics is often regarded as an exponential function of time (Day 1974). Here the delay varies with time t as $\exp(kt)$

$$\tau = \tau_0 e^{kt} .$$

Hence, according to (69)

$$\tau_* / \tau_0 = e$$

and, as at the critical point the delay is equal to the argument value,

$$t_* / t_0 = e .$$

This agrees with (69).

Therefore, the sequence of critical argument values can be represented as

$$t^k = e t^{k-1} , \quad k = 0, 1, 2, \ldots \tag{70}$$

where k is the number of a critical point.

If the argument is the system's age, t^k is called a critical boundary or critical age. Thus, for exponential development processes the critical argument values are arranged in such a way that their ratio is constant and equal to the Napierian number e, the base of natural logarithms.

The Napierian number is well known and widely used in the theory and practice of mathematical modeling. Thus, for example, when growth is modeled using the Bertalanffy and Gomperz equations, the ratio of the limiting size to the size at the point of curve inflection is equal to e (Mina and Klevezal 1976). The consistency of the models seems to be explained by the fact that the critical size ratio is defined by the form of the relationship used.

A time constant is defined as the time required for a characteristic to fall to $1/e$ of its initial value (Ginkin 1962).

The highest noise immunity in information theory is known to be achieved when probability is $1/e$ (Shennon 1963). This result conforms with the general properties of power-exponential functions like entropy (Savelov 1960).

Another area where the number e acts as a critical constant is the building of hierarchical structures in many identical components. A hierarchy with the module e is perfect in terms of information transmission time (Fleishman 1971). In the integer case this corresponds to a ternary structure widely used in linear hierarchies (Ober-Krie 1973).

For the exponential process of unstable type, we have at the critical point $t_* = \tau$

$$\theta_* = t_* / t_0 = \frac{1.293}{3/2\,\pi}$$

or, according to (28, 31, 32),

$$\frac{1.293}{3/2\,\pi} = e^{-1.293} ,$$

from which

$$\theta_* = t_*/t_0 = e^{-1.293} = 1/3.644 . \quad (71)$$

It means that for the exponential process of unstable type the ratio of ages or argument values at two sequential critical points is 3.644.

The total range for two processes of stable and unstable types accounts for

$$t_{1*}/t_0 = e^{2.293} = 9.95 . \quad (72)$$

Thus, for two processes of stable and unstable types, the full change in ages at the critical points is about an order of magnitude.

3.4 Hierarchy of Critical Constants

Chapter 3 has covered the hierarchy of development processes. The sequence of hierarchical level models has been built by introducing the dependences of growth rate or relative growth on time as an independent variable in the models. When the growth rate is invariable or varies in inverse proportion to t or $t \ln t$, we get an exponential mode, an allometric mode, or the envelope of allometric modes, respectively.

The exponential mode describes the nature of development on the lower level; the allometric mode comes as the envelope of exponential processes when the growth rate decreases. A series of allometric models is aggregated in much the same way.

As shown in Chap. 2, the development equation (4) allows the system characteristics to vary exponentially in the domain defined by

$$u = \frac{\alpha}{\tau} \quad (73)$$

where $\alpha = 1.293$ or -1.0 for the processes of unstable and stable types, respectively. Therefore, at the vicinity of critical boundary (73), the development process is defined by

$$\dot{x} = \frac{\alpha}{\tau} x .$$

As a result, the aggregation is made in the above manner, because $\tau = b_1 t$ for the exponential processes with falling growth rates. Hence,

$$\dot{x} = \frac{\alpha}{b_1 t} x = \frac{B}{t} x$$

and we get an allometric mode. A series of allometric modes involves $\tau = b_2 t \ln t$, and their envelope is derived as a description of system behavior in the vicinity of the critical boundary

$$\frac{dx}{dt} = \frac{\alpha}{b_2 t \ln t} x \ .$$

This means that delay on the previous level generates a function K(t) for the development equation of the next level. Thus, the hierarchy of development models, in which the uniform process counts as the first level, has the form

$$\begin{aligned} \frac{dx_2}{dt} &= k x_2 (t - b_2 t) \\ \frac{dx_3}{dt} &= \frac{B_3}{t} x_3 (t - b_3 t \ln t) \\ \frac{dx_4}{dt} &= \frac{B_4}{t \ln t} x_4 (t - b_4 t \ln \ln t) \end{aligned} \tag{74}$$

where x_i, $i = 1, 2 \ldots$, is the system size on the ith hierarchical level.

Let time be measured in proportion to the length of the primary process on the lower hierarchical level:

$$\theta = t/t_0 \ .$$

Then, in line with the function of t used as an argument of the development equation on each hierarchical level, there exists a bound on the relative age of the developing system; after the bound is reached, the corresponding level is activated. Thus, any exponential process begins at zero time. An allometric process is activated at $\theta = 1$ and the envelope of allometric processes ($\theta_4 = \ln \ln \theta$) is activated at $\theta = e$. The next development level is activated at $\theta \approx e^e$, and so on.

Notice that the activation of a hierarchical level takes place at the value of dimensionless time which is defined by the ratio of ages or arguments at the critical points on the level that is two steps below the level in question. Thus, the e ratio of the sequential critical ages is characteristic of the exponential processes (68) and it activates the process which is the envelope of allometric development modes. The e^e ratio (56), which is characteristic of the allometric processes of stable type, activates the level that is two steps above the allometric level (Fig. 26).

Thus, the hierarchy of structural levels is observed in system development. The hierarchy for stable processes is characterized by critical ratios which are defined by the following recurrent relationship

$$N^k = e^{N^{k-1}} \ , \qquad k = 0, 1, 2, \ldots \tag{75}$$

where N^k is the ratio of ages or arguments at two sequential critical points. The argument values are called critical constants.

If the process of stable type has $N^0 = 0$, then the ratios for different critical levels are listed in Table 2.

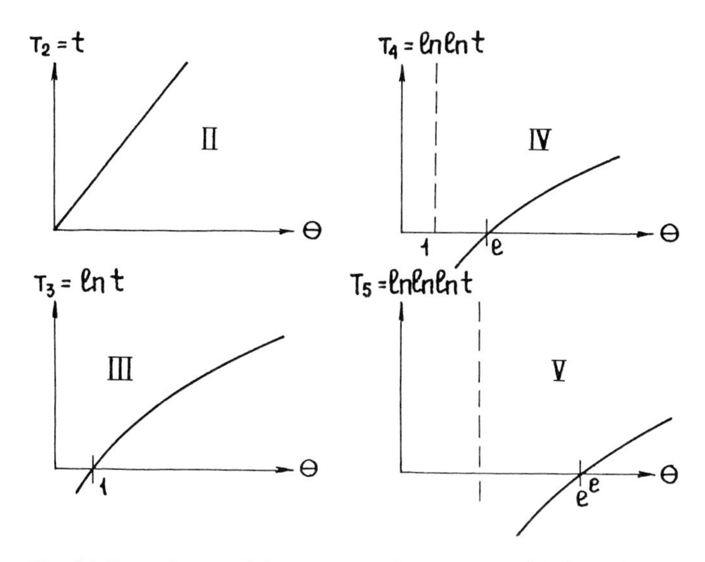

Fig. 26. Dependences of the arguments for processes with decreasing growth rates on dimensionless time for different hierarchical levels. The points of transition to another hierarchical level are defined by the value of dimensionless time T = 0. *II* exponential mode; *III* allometric mode; *IV* envelope of allometric modes

Table 2. Critical constants for the processes of stable type and the relational ages when the hierarchical levels are activated

Number of the hierarchical levels, k	Hierarchical levels activated by the ratio	Ratio of arguments at the sequential points, N^k	Name of a hierarchical level
0	2	0	
1	3	1	
2	4	e	Exponent
3	5	e^e	Allometry
4	6	$e^{(e)}$	The envelope of allometric modes

In accordance with the relationships (4) and (1), for the allometric and exponential processes of unstable type, the critical constants make their appearance on the exponent level and are defined by the following initial ratio.

$$N^0 = e^{-1.293} .$$

Thus, the entire life cycle is divided into stages. Transitions between stages proceed in an abrupt manner. This is associated with the abrupt change in the rate of exponential growth, in allometry parameter or in the parameter of allometry envelope, and so on. The change points are called critical. At these points the behavior of a developing system undergoes a qualitative change. The critical points are arranged in sequence by their significance in the hierarchy. The significance of a critical point to the development process rises as the difference in age and size between the sequential critical points increases, i.e., as the hierarchical level becomes higher in the sequence: exponent, allometry, the envelope of allometric modes, and so on.

The large critical ratios are of particular interest; they are well exhibited in the experimental data. It is interesting that the estimate of the critical constant of stable type for the envelope of allometric modes was obtained by Schmalhausen (1984) from experimental data. He examined an organism's life cycle from the beginning of embryonic development to its death as the sequence of allometric modes and found out that the largest volume of any organism is dependent on its initial size. Schmalhausen introduced an index which he called a specific growth capacity

$$u = \int_1^t \frac{B(t)}{t} dt$$

where B is the allometry parameter.

In the case of allometric mode a change in volume is defined as

$$\frac{dv}{v\,dt} = \frac{B}{t} \quad \text{or} \quad \ln v = \int_1^t \frac{B}{t} dt + \ln C_1 \ , \quad \frac{v}{v_0} = e^u \ . \tag{76}$$

Because the entire development process includes a number of allometric modes (see Fig. 1), we can find the specific growth capacity by integrating the sequential allometric modes, with the values of allometric parameters being constant. Then

$$u = \sum_{i=1}^{n} B_i \ln \frac{T_i}{T_{i-1}} \ .$$

The specific growth capacities calculated by Schmalhausen (1929, 1984) are listed in Table 3.

A conclusion has been drawn from the data that the value of 20–25 is the maximum specific capacity which can be achieved by the unity of the living substance in the differentiated organism.

From (76), the ratio of the limiting volume to the initial volume is

Table 3. The specific capacities of animal growth (According to Schmalhausen 1929, 1984)

Species	Specific growth capacity		Total
	Embryonic growth	Postembryonic growth	
Starred sturgeon	2.2	17.2	19.4
Pike	1.8	15.3	17.1
Hen	9.5	3.8	13.3
Mouse	8.9	2.3	11.2
Rat	13.7	4.3	18.0
Guinea pig	13.6	2.3	15.9
Pig	15.7	4.9	20.6
Man	21.0	3.3	24.3

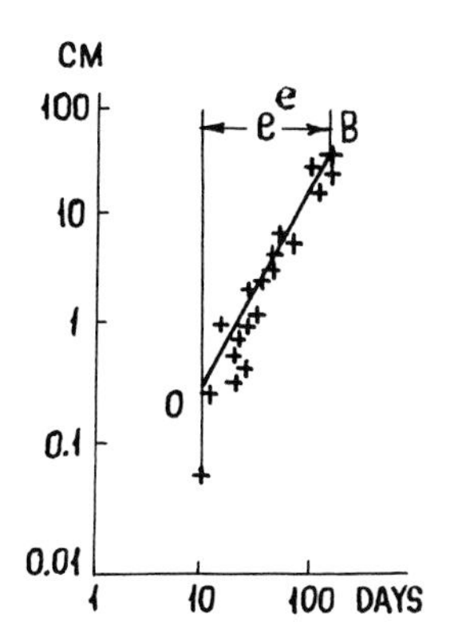

Fig. 27. Age dynamics of the linear size of the pig embryo. *Abscissa* age; *Ordinate* size, *0* beginning of organismic development; *B* birth (Data from *Biology Data Book* 1964)

$$V_*/V_0 = e^{20} - e^{25} = 10^{8.7} - 10^{10.9} \ .$$

The ratio is somewhat overestimated, since in a strict sense, the preorganismic stages of embryonic development are not allometric, and then it is not valid to include them in the calculation of specific growth capacity. The specific growth capacity, as a development characteristic, becomes true from the beginning of organismic development (Fig. 27), when the first allometric mode starts. This has mostly affected Schmalhausen's data concerning the development of large animals for which preorganismic stages are longer than those for small animals. Man's organismic stages begin at the 19th day (Stanek 1977). Then the recalculation of data in Table 3 yields the specific growth capacity of 13.2. A pig's organismic development begins at the 10th day, which coincides with the beginning of the first allometric mode (*Biology Data Book* 1964, see Fig. 27). In this case the specific growth capacity is also equal to 13.2. It means that the specific growth capacity for the whole development period accounts for 13–16. The ratio of the limiting volume to the initial one is then estimated as follows.

$$V_*/V_0 = e^{13} - e^{16} = 10^{5.6} - 10^{7} \ ,$$

which agrees with the critical ratio of hierarchical level 4 (Table 2).

It is interesting that Archimedes (1962), in his days, introduced a power-exponential sequence which is generated like the sequence (75) but the base equals 10^8

$$N^k = (N^{k-1})^{10^8} \ .$$

The sequence (76) represents an exponential series. Just as in the continuous case, where the development becomes allometric at certain points because of the action of memory mechanisms, so for the discrete mechanisms for forming critical points the allometric relationships will also hold true. That is, the following sequence of critical constants can be expected to come into being

$$N^k = (N^{k-1})^{\alpha} \ .$$

It is necessary to specify an initial term of the series to obtain the values of critical constants from it. Allometry comes up as the envelope of exponential modes, so the critical ratio obtained for the exponential process can be adopted for the initial value. On the other hand, the critical ratio for the allometric process must be valid, which allows the exponent of a power to be specified. Then provided $X^0 = e$ and $X^1 = e^e$, we found that $\alpha = e$, from which

$$N^k = (N^{k-1})^e$$

and the third term of the sequence is

$$(e^e)^e = e^{(e^2)} = 1618.173\ldots\ .$$

In the past, the value was regarded as the largest number ($40 \times 40 = 1600$).

The first two terms (e and e^e) of critical constants of exponential and allometric modes turn out to be the same, but then the relationship for the exponential mode gives considerably larger values. The processes under consideration require a time scale to be contracted, which is characteristic of falling growth rates.

In contrast, the time scale is expanded at the intervals where the growth rates increase. In this connection, the law of forming critical constants with level shifts appears in a different way. Here the time scale changes from linear to exponential. So the critical constants are formed by the function that is the inverse of the exponent function, i.e., by the logarithmic function. Hence,

$$N^k = \ln N^{k-1}$$

or

$$N^{k-1} = e^{-N^k}, \quad k = 0,\ -1,\ -2,\ -3,\ldots\ .$$

The general law of forming critical constants which holds for both classes of processes is written as

$$N^{(k)} = e^{(\mathrm{sign}\ k)N^{|k|-1}}, \quad k = 0,\ \pm 1,\ \pm 2,\ldots$$

with $N^0 = 0$ or $e^{-1.29}$ the initial conditions for the processes of stable and unstable types, respectively. The values of critical constants for the processes of stable and unstable types are listed in Table 4.

Table 4. Critical constants for different hierarchical levels

Level k	Stable type		Unstable type	
	N^k	$(N^k)^{-1}$	N^k	$(N^k)^{-1}$
∞	0.5671	1.7632	0.5671	1.7632
−6	0.5454	1.8335	0.5566	1.7967
−5	0.6062	1.6495	0.5860	1.7066
−4	0.5005	1.9981	0.5345	1.8710
−3	0.6922	1.4447	0.6265	1.5963
−2	1/e	e	0.4677	2.1383
−1	1	1	0.7600	1.3158
0	0		0.2744	3.6442
1	1	1	1.3158	0.7600
2	e	1/e	3.7276	0.2683
3	e^e	$(e^e)^{-1}$	41.5792	0.0241
4	e^{e^e}	$(e^{(e^e)})$	1.1420×10^{18}	

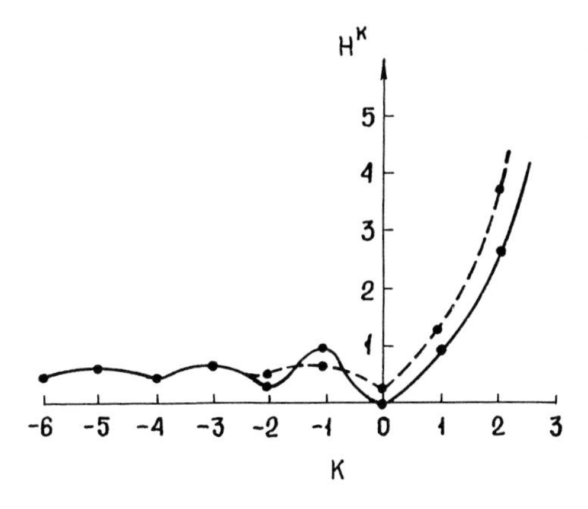

Fig. 28. Critical constants for different hierarchical levels. The critical points for processes of stable and unstable types are connected by the *solid* and *dashed curves*, respectively

Table 5. Comparison of the golden section gnomon with critical constants

Golden section gnomon	1	1.128	1.272	1.435	1.618	1.825	2.058
N^k	1		$1.274 = \sqrt[4]{e}$	1.445	1.649	1.834	1.998
Level k	-1		-2	-3	-5	-6	-4

Some critical constants listed in Table 4 are known and widely used. Thus, for example, the critical constant $1/e$ of level -2 is the inverse of the critical constant of level 2. The importance of the Napierian number e has been discussed above.

The critical constant $(1/e)^{1/e}$ of level -3 is close to $\sqrt{2} = 1.414$ which is widely used in proportion theory (Rybakov 1984). Hegel (1986), in his thesis *De orbitis planetarum* (published in 1801), proposed to take $\sqrt[3]{3} = 1.442$ as the origin of the planet sequence. The number is very close to $(1/e)^{1/e}$.

The critical ratio of level -4 defines the twofold change of the time scale, which corresponds to one octave as the basis of the musical scale. Feigenbaum (1980) believed that doubling the period is characteristic of a system transition from simple periodic motion to a complex nonperiodic one. He gives a great deal of examples where such a development mechanism takes place.

Figure 28 shows the critical constants for different hierarchical levels. On the levels below -3, the constants for the processes of stable and unstable types are seen to be practically identical and tend to a single limit. As the zero level is approached from the left, the magnitude of critical constant deviation gains, and on passing to the positive level numbers the sharp growth of the critical constants is observed.

In ancient Greece, there existed a golden section gnomon which allowed an octave to be divided in the golden ratio. The markers of the gnomon, in comparison with the critical constants from Table 4 for the stable type of processes, are listed in Table 5. As seen, the first six constants have their analogs on the gnomon.

The approximate value of the asymptotic critical constant ($0.5671 = 1/1.76325$) looks like $\sqrt{3} = 1.7320$ in proportion theory (Rybakov 1984).

3.5 Synchronization of Critical Boundaries of Different Hierarchical Levels

In the previous section we have discussed the hierarchy of critical ratios for processes which take place on different hierarchical levels. The intensity of a critical phenomenon can be expected to increase substantially in those cases where some critical boundaries of different hierarchical levels prove to be synchronized.

We begin by establishing the synchronization condition for the processes of levels 1 and 2. Level 1 is distinguished by constant growth rates. These are, for example, an interval between sequential cell divisions, which serves as a metronome for developing cell populations, a sequence of environmental cycles which defines the development in the biosphere, geosphere, and so on. The equal time intervals specify a chain reaction which is described as an exponential process. For the exponential process, a sequence of critical argument values is defined by the expression (70)

$$t^k = e t^{k-1} \ , \quad k = 0, 1, 2, \ldots$$

where t^k is the argument value at the critical point k. In the case where the argument is a system age, this value is a critical age or boundary.

We will find the condition that a multiple of the boundaries or the equal time interval of the first level fall within the range between the time moments defined by (70).

Assume that t^k is the age associated with one of the unevenly spaced boundaries and that the range between t^{k-1} and t^k contains strictly n equal time intervals T^k. By (70), the condition is written as follows

$$t^k - t^{k-1} = (1 - 1/e) t^k = n T^k \ . \tag{77}$$

Assume that one time interval T^k lies outside the range (t^{k-1}, t^k), i.e., the age $t^{k-1} - T^k$ falls at some unevenly spaced boundary $t^{k-\gamma}$

$$t^{k-1} - T^k = (n-1) T^k \ .$$

Then

$$t^k e^{-1} - t^k e^{-\gamma} = (n-1) T^k \ . \tag{78}$$

Combining (77) and (78), we get a set of simultaneous equations

$$t^k (e-1) = e n T^k$$

$$t^k (e^{\gamma-1} - 1) = e^{\gamma} (n-1) T^k \ . \tag{79}$$

Hence

$$\frac{e-1}{e^{\gamma-1}-1} = \frac{e}{e^{\gamma}} \frac{n}{n-1}$$

or

$$\frac{n}{n-1} = \frac{e^{\gamma-1}(e-1)}{e^{\gamma-1}-1}$$

and

$$n = \frac{e^{\gamma-1}(e-1)}{(e^{\gamma-1}-1)\left(\frac{e^{\gamma-1}(e-1)}{e^{\gamma-1}-1}-1\right)} = \frac{e^{\gamma-1}(e-1)}{e^{\gamma}-2e^{\gamma-1}+1} .$$

Then

$$n = \frac{e-1}{e-2+e^{1-\gamma}}$$

and

$$\lim_{\gamma\to\infty} n = \frac{e-1}{e-2} = 2.39 .$$

So the integer values of n may be only n = 2 or n = 1. We will find the integer value of n.

By convention, γ not less than two is an integer. When $\gamma = 2$

$$n = \frac{e-1}{e-2+e^{-1}} = 1.58$$

while at $\gamma = 3$

$$n = \frac{e-1}{e-2+e^{-2}} = 2.01 . \tag{80}$$

Further increasing γ up to ∞ leads to no occurrence of a new integer value of n. So the only value of γ is $\gamma = 3$, yielding n = 2.

The requirements for synchronization across m critical levels yields the following relationship

$$t^k = n \frac{1}{1-e^{-m}} T^k .$$

Table 6 lists the ratio $t^k/(nT^k)$ of the boundary age to time interval T^k as a function of m.

As follows from Table 6, it is advisable to seek synchronization boundary points for m = 1 (the above mentioned case of synchronization between neighboring boundaries) and m = 2 (the synchronization of each second boundary), since the rest of ratios are associated with the boundary points specified by the length of time interval T^k because of

$$\lim_{m\to\infty} t^k = nT^k .$$

Table 6. Ratio of the critical boundary age to time interval T^k as a function of the synchronization range

m	1	2	3	∞
$t^k/(nT^k)$	1.58	1.15	1.05	1

Consequently, the complete set of critical boundaries where the first and second levels are synchronized is defined as follows.

For m = 1, the basic synchronized boundaries (up to the order of k−3) are

$$t^k = \frac{2}{1-1/e}\,T^k = \frac{2e}{e-1}\,T^k \tag{81}$$

and the intermediate nonsynchronized boundaries, depending on the time interval T^k, are

$$t^k - T^k = \left(\frac{2}{1-1/e} - 1\right) T^k = \frac{e+1}{e-1}\,T^k \ .$$

For m = 2,

$$t^k = \frac{n}{1-e^{-2}}\,T^k = n\,\frac{e^2}{e^2-1}\,T^k \ .$$

It is easy to show that synchronization with more distant boundaries is impossible. Hence, the case m = 1 is of interest because three of the unevenly spaced, critical boundaries prove to be synchronized with the equidistant boundaries of the first level.

By (81),

$$t^k = 2e(e-1)^{-1}T^k = 3.16T^k \ .$$

Given the time interval T^k, the corresponding critical boundary points under the above synchronization conditions can be determined. Then it appears that two equal intervals fall between two sequential, unevenly spaced critical boundaries, and before them lies one more time interval T^k (Fig. 29), which is synchronized with the boundary t^{k-3}. The critical boundary points are presented in Table 7.

Combining (70) and (81), we find

$$T^k = eT^{k-1} \ . \tag{82}$$

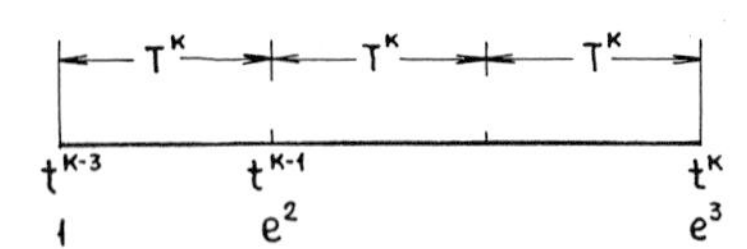

Fig. 29. Synchronization of equidistant and unevenly spaced boundaries

Table 7. The critical boundary points for the synchronization range of the first and second levels

Name	Designation	Numerical expression
Basic k	t^k	$t^k = 3.16\,T^k$
Intermediate k		$t^k - T^k = 2.16\,T^k$
Basic k − 1	t^{k-1}	$t^k - 2\,T^k = 1.16\,T^k$
Basic k − 3	t^{k-3}	$t^k - 3\,T^k = 0.16\,T^k$

It means that the time interval T^k is a geometric progression with the common ratio of e. Then the ratio of ages of the synchronized, unevenly spaced boundaries coincides with that of the time intervals T^k

$$t^k/t^{k-1} = T^k/T^{k-1} = e \ .$$

Thus, the total range where the equal intervals are synchronized with the unevenly spaced boundaries of the stable exponential type does not exceed the $1 : e^3 = 1 : (20.085)$ ration.

Because the synchronization range of equal time intervals covers process characteristics of up to the $(k-3)^{th}$ order, one can expect the bound of a time interval at least of the $(k-1)^{th}$ order to appear between the unevenly spaced, critical boundaries t^k and t^{k-1}.

According to (77), the equal time intervals are defined by Eq. (82). The equidistant boundaries located between t^k and t^{k-1} yield the following collection of numerical expressions:

$$\begin{aligned} t^{k-1} + T^{k-1} &= 4.16\,T^{k-1} = 1.53\,T^k \\ t^{k-1} + 2\,T^{k-1} &= 5.16\,T^{k-1} = 1.90\,T^k \\ t^{k-1} + 3\,T^{k-1} &= 6.16\,T^{k-1} = 2.27\,T^k \\ t^{k-1} + 4\,T^{k-1} &= 7.16\,T^{k-1} = 2.63\,T^k \\ t^{k-1} + 5\,T^{k-1} &= 8.16\,T^{k-1} = 3.00\,T^k \ . \end{aligned}$$

As shown in Sects. 3.2 and 3.3, the allometric development, i.e., the process of the third hierarchical level involves the critical ratios (56) and (64), which specify the $1 : 19.943$ ratio of ages at two sequential critical points for the stable and unstable process types. As a result, the pair of the allometric modes of stable and unstable types proves to be practically synchronized with the synchronization range of the first and second levels. The ratio (56) for the allometric process of stable type which is equal to e^e specifies the interval of stable development, and the interval from e^e to the end of the synchronization range [defined by Eq. (64) for the allometric process of unstable type] specifies the system's reorganization phase.

We find the fraction of the reorganization phase in the whole synchronization range as

$$\frac{e^3 - e^e}{e^3}\,100 = 24.6\% \ .$$

The large synchronization period may be thought to comprise the allometric phase which takes three quarters of its length, and the reorganization phase whose length is a quarter of the period. The location of the reorganization phase in the sequence of the unevenly spaced, critical boundaries is obtained as follows:

$$t^k - 0.246\,(t^k) = 0.754 \cdot 3.16\,T^k = 2.38\,T^k \ .$$

Finally, we will take into account the synchronization with the critical range of the exponential processes of stable and unstable types. To this end, we take advantage of the critical ratios (68) and (71), and their product (72) which practical-

Table 8. The location of boundary points within the synchronization range

No.	Boundary point	Age expressed in terms of T^k	No.	Boundary point	Age expressed in terms of T^k
1	Basic k – 1	1.16	6	Equidistant k – 1	2.27
2	Unstable exponential	1.56	7	Reorganization phase	2.38
3	Equidistant k – 1	1.53	8	Equidistant k – 1	2.63
4	Equidistant k – 1	1.90	9	Equidistant k – 1	3.00
5	Equidistant k	2.16	10	Basic k	3.16

ly accounts for a half (9.905) of the synchronization range. We find the fraction of the pair of the exponential processes in the synchronization range as

$$\frac{e^3 - 9.905}{e^3} 100 = 50.69\%$$

and their location in the sequence of the unevenly spaced boundaries is obtained as follows:

$$t^k - 0.507(t^k) = 0.493 \cdot 3 \cdot 16 T^k = 1.56 T^k \ .$$

Thus, the synchronization range includes the unevenly spaced boundaries of the following processes: stable exponential, stable and unstable exponential, stable and unstable allometric processes – with the latter forming the reorganization phase – as well as a sequence of equal time intervals of the kth and $(k-1)^{th}$ orders. These boundaries are listed in Table 8.

We call the boundary set a development unit or development link. To obtain the location of the development unit in the real process it is necessary to know the length of time interval T^k or the location of a certain well-known critical boundary. In the second case, some difficulties may arise, which are related to the uncertainty about the reference point.

The basis of the development unit is formed by a vigesimal (based on the number 20) system ($e^3 = 20.085$), which was used by the people of many countries. The basis of the system is associated with the number of fingers and toes of a man's hands and feet (Yushkevich 1970). Notice that the ratio of the lengths of time intervals of the kth and $(k-3)^{th}$ orders equals

$$T^k : T^{k-3} = e^3 \ .$$

There are three equal time intervals of the kth order in the development unit, so the total number of time intervals of the $(k-3)^{th}$ order is equal to $3e^3 = 60.2$. This is the basis of the sexagestimal system (based on the number 60), which is widely used (60-year cycles in Eastern calendars, an hour equals 60 minutes, a minute equals 60 seconds).

The synchronization of the level 2 and 3 critical constants (e and e^e) is often corroborated by experimental data. In the following we will illustrate such synchronizations. They are based on the following relationship

$$e^n = (e^e)^m \quad \text{or} \quad e^{n/m} = e^e$$

where n/m is the fraction approximating to e (Khinchin 1961). The approximating fractions for e are represented by a series: 2/1, 3/1, 8/3, 11/4, 19/7, 106/39... It is the fractions that are widely used, because the consequent fractions are expressed in terms of large numbers. Hence, in the series of critical points, one can expect more intensive phenomena in the vicinity of $e^8 = e^{e^3}$, $e^{11} = e^{e^4}$, $e^{19} = e^{e^7}$.

The synchronization of critical boundaries of each hierarchical level and the level interaction mechanism requires additional detailed studies.

Pay attention to the fact that the boundaries of the development unit are related to Feigenbaum's constant (1980)

$$F = \frac{\lambda_{n+1} - \lambda_n}{\lambda_{n+2} - \lambda_{n+1}} = 4.6692016$$

where λ_n is the parameter value at which a period is doubled for the nth time. Two basic boundaries (e^3 and e^2) of the development unit will be regarded as critical parameters. Among all the critical boundaries we will find a point where the following equation

$$\frac{e^3 - e^2}{e^2 - b} = b$$

holds true.

Solving the equation for b, we find its roots $b_1 = e$ and $b_2 = e(e-1) = 4.670774$. The second value is different from Feigenbaum's constant, with the absolute error being 0.000337. The development processes can be expected to be characterized by other constants, which are defined from the critical points in a nonmultiplicative manner.

The results obtained allow the development processes to be treated in a unified manner, with differences in the formation of system characteristics on the different hierarchical levels taken into account. The critical constants obtained by considering the intervals of stable development manifest themselves in system characteristics involving transitions to new hierarchical levels. The critical phenomena in the development processes can be classified according to their significance and the hierarchical level occupied.

The following sections are devoted to extending the results based on the data concerning the development as some natural systems.

4 Critical Levels of Temporal Characteristics of Systems

In the preceding section, we have obtained the hierarchy of critical constants and, on the basis of this hierarchy, a cell of development. All the following results are connected with an analysis of manifestations of these critical constants, either separately or in their interrelation (for example, in a cell of development).

The main features of any system are associated with their temporal and spatial characteristics and their structure. Therefore, we describe each of these characteristics in separate sections.

When considering critical levels of temporal characteristics of systems, we encounter the basic problem whether there exist general rhythms which form systems of different structural levels. To answer this question we should know the regularities of formation of natural rhythms and consider them in the analysis of data on the development of systems of different structural levels.

A study of the duration of rhythms of galactic processes has shown that these processes form a succession represented by a geometric progression with the module e. Further, proceeding by hierarchy, the solar system shows a succession of periods of planet rotation determined by even periods of Sun rotation, through the cell of development (except the Earth and Neptune). Synchronizations of rotations of the Earth and Moon, and of the Sun and Moon, recorded in lunar and solar-lunar calendars, are based on the critical constant e.

4.1 Periodization of Critical Boundaries in the Development of the Earth's Crust

In Sect. 3.5 we have obtained ratios characterizing a sequence of boundaries represented by synchronized boundaries of an exponential type, e and e^3, repetitions of cycles within the framework of arithmetical progression as well as by critical levels of an allometric type. The general structure of a standard element of periodization is given in Table 7.

An envelope curve of successive exponential regimes of development at falling rates within prolonged time intervals has an allometric nature. We shall consider experimental data in double logarithmic coordinates providing a linear character of allometric dependence. Then, a relation e^e between consecutive critical ages will be found in the range where the allometric curve occurs. Further, up to the range of synchronization of exponential boundaries (e^3) a reconstruction phase will take place.

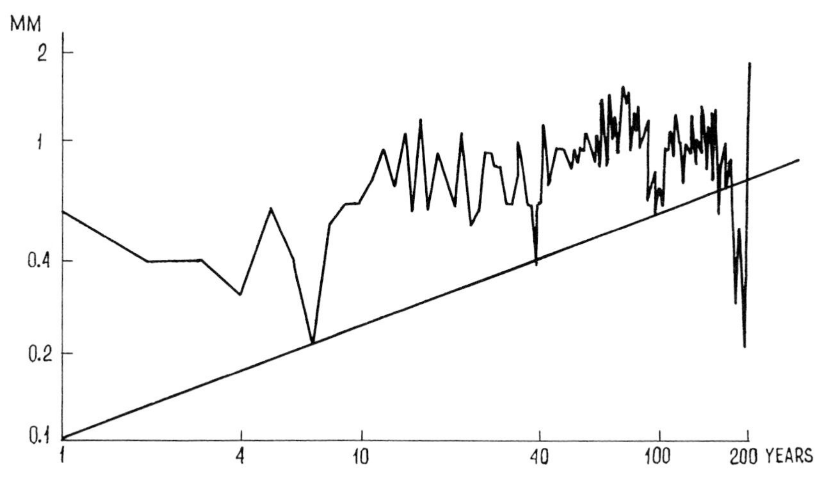

Fig. 30. Changes in the width of annual rings on cross section of a Japanese cypress for the first 200 years of growth. *Abscissa* age; *Ordinate* annual increments (Data from Arakawa 1954)

Let us deal with the data of K. Yamasawa on a change in the width of growth rings on the cross section of a 1000-year old Japanese cypress for the first 200 years of growth (Arakawa 1954). One can see in Fig. 30 a series of consecutive cycles of change in the width of growth rings, the lower boundary for which is an allometric curve to the 160th year of growth. Further, a transition process is observed in the area below the allometric dependence; and then a rapid growth characterizing the start of a new stage of development occurs. Thus, the whole cycle of development consists of fluctuations above the allometric dependence (we call this curve portion an allometric phase) and a sharp reduction of increments to the minimum (reconstruction phase).

Figure 30 shows that the cycle ends in the 195th year, the scale being marked off from the start of the cypress growth. Let us regard this boundary as corresponding to that of synchronization e^3. Then, the calculated duration of the reconstruction phase will be, according to Sect. 3.5:

$$195 \cdot (e^3 - e^e)/e^3 = 48 \text{ years.}$$

Thus, the consecutive boundaries will fall on $195 - 48 = 147^{th}$ year and, following the same cycle, on the 99th, 51st and 3rd year. Meanwhile, measurements showed the last increments in the 99th, 39th and 7th year of growth.

This example is, in a certain sense, unique in the volume of information about developing system. Usually, we have far more moderate evidence on the results of development of biological systems.

Gnezdilova et al. (1976) give evidence on a change of egg volumes during ovogenesis of the sea urchin *Strongylocentrotus nudus* (see Fig. 2). Development of the egg cell lasts from September to February and, in the given case, the complete period of synchronization is 5.8 months. Hence, duration of the reconstruction phase is of the order of

$$5.8 \cdot (e^3 - e^e)/e^3 = 1.4 \text{ months.}$$

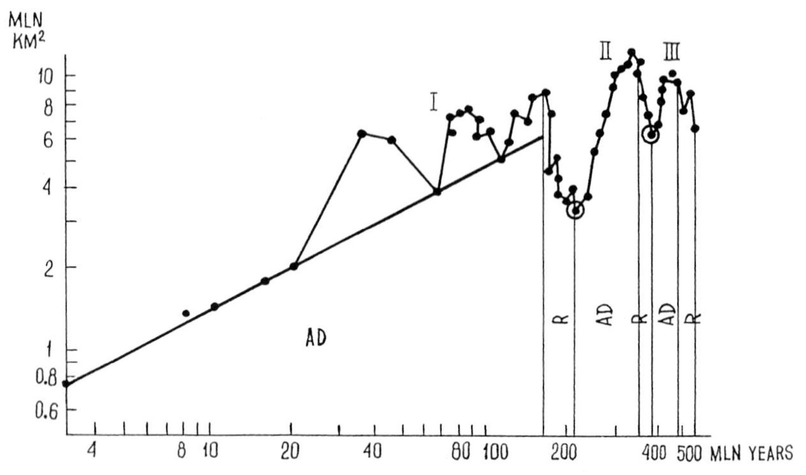

Fig. 31. Changes of areas occupied by sea deposits at the Phanerozoic on USSR territory. *Abscissa* time; *Ordinate* areas. Logarithmic scale used for both axes. *I* Alpine, *II* Hercynian and *III* Caledonian cycles of orogenesis. The *inclined curve* shows a tendency of development of one of the *I* allometric phases. *Circled points* denote transition from one cycle of development to another one. Periods of allometric development (*AD*) alternate with periods of reconstruction (*R*) (Data from Karogodin 1975)

So, the end of allometric development should fall on $5.8 - 1.4 = 4.4^{\text{th}}$ month, which is the actually observed time.

The previous boundaries are: 3, 1.6, and 0.2 months. Of these, the point at the age of about 3 months is on the allometric curve, which must be an envelope curve of minimal values. Gnezdilova et al. (1976) note that a cycle of development consists of three consecutive exponential portions and one reconstruction cycle that corresponds in duration to the regime of exponential growth.

Measurement of the areas occupied at different geological times by sea deposits on USSR territory has given similar results (Karogodin 1980). By fluctuations of these areas one can appreciate cycles of transgressions and regressions (Fig. 31) – there are three in the figure. Time of start, end, and duration of phases is given (Table 9).

Each cycle consists of an allometric phase (the tendency, corresponding to an enveloping minima dependence) and a reconstruction phase (portion of drop down). The reconstruction phase is followed by another allometric phase of a new cycle which terminates in a reconstruction phase. Transition from one cycle to another occurs in the stage of maximal regression. If we take the end of the cycle in each case for the beginning of the next cycle, we shall obtain dependencies as shown in Fig. 32; hence, it appears that characteristics of cycles repeat. According to the data of Table 9, reconstruction phases occupy 21%, 17% and 24% of the general cycle duration, which corresponds to the theoretical evaluation of the reconstruction phase share in the whole time of the synchronized range of the development process.

For geological processes, the position of certain boundaries is determined with a high level of reliability; they can be classified by their relative significance in the geological history of the Earth.

Table 9. Phases of transgression and regression cycles

Absolute time, million years	Phase	Duration, million years
11		
	Allometric	155
165		
	Reconstructional	45
210		
	Allometric	150
360		
	Reconstructional	30
390		
	Allometric	130
520		
	Reconstructional	40
560		

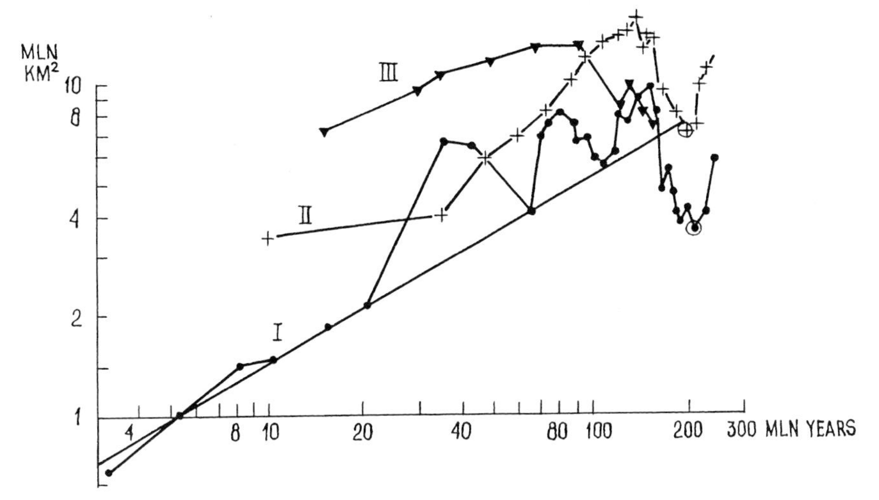

Fig. 32. Matching of cycles in the development of the Earth's crust (*I, II, III*) presented in Fig. 31. Designations as in Fig. 31

Geological boundaries. Many researchers note that geological history includes stages of comparatively calm development and stages of more active tectonic movements and foldings between the former. The latter are taken as boundaries between geotectonic cycles. These boundaries are distinguished against the general trend in the development of the Earth's crust, expressed by a considerable reduction in the degree of metamorphism. This is most clearly seen, when comparing rocks of Archaean, Baikal, and Cainozoic geosynclines, in a decreasing manifestation of granitization, as well as in a decrease of the area occupied by geosynclines on continents (Ronov 1976).

Boundaries dividing basic stages in the development of the Earth's crust may be taken as the largest in geological history. Belousov (1975) distinguishes permobile unstable protogeosynclinal, and stable geosynclinal-platform stages. The first stage, approximately coinciding with the Archaean, is characterized by a general development of the geosynclinal regime. It came to an end 2.8–3 billion years ago. The second stage, coinciding with the early and middle Proterozoic, is characterized by the appearance of unstable platforms of a type of median masses and linear geosynclines. It came to an end 1.6–1.8 billion years ago. The third stage is distinguished by the appearance of stable platforms. It can be divided again into two stages. The first one can be provisionally called a mobile geosynclinal-platform stage during which eugeosynclines went on appearing within platforms and aulakogenes were widely spread. It corresponds to the late Proterozoic. During the second mobile geosynclinal-platform stage within ancient platforms, eugeosynclines ceased to appear and aulakogenes got a subordinate spread. From that time platforms began to occupy more than half the area of continents. The boundary between these stages on ancient platforms is approximately on the level of 570 million years. It coincides with the boundary between aulakogene and plate stages distinguished by Khain (1973).

All these four stages ended with epochs of folding, granitization, and large tectonic reconstructions. They coincide approximately with Saam-Belozyorsk, Svekofeno-Karelian, and Baikalian foldings. It is known that epochs of tectonic activation and foldings occupy comparatively long time intervals. We take the time of the maximal manifestation of granitization, coinciding mainly with terminal stages of folding, as conditional, shorter moments for the Precambrian.

The time of ending of the above considered stages was different for each continent. Belousov (1978) points out that on southern continents the unstable protogeosynclinal stage ended later than it did on northern ones, about 0.6 billion years ago. On epihercynian plates, the mobile geosynclinal-platform stage ended not in the early Paleozoic but the late Triassic–early Jurassic, 0.2 billion years ago. However, the ending of stages always coincides with some period in geological history. For the first example, this is the Baikal folding, for the second one the ending of the Hercynian folding.

It should be also mentioned that change in the character and intensity of the manifestation of tectonic processes from stage to stage corresponds to a general tendency in the development of the Earth's crust, that is, to a general reduction of tectonic activity.

Within the most important boundaries of geological history, dividing stages in the development of the Earth's crust, there exists a system of less significant boundaries. The number of the largest ones is usually two or three within each stage. At a permobile stage, a boundary sometimes appears on the 4-billion-year level. More distinct and generally accepted is the boundary at 3.5–3.6 billion years. The first significant epoch of granitization is connected with this boundary (Table 10).

Within the unstable protogeosynclinal stage, a boundary on the 2.5–2.6 billion-year-level is distinct. A wide-ranging introduction of granitic intrusions coincides with it. A boundary at 1.9–2.1 billion years is also generally accepted. The mobile geosynclinal-platform stage includes two distinct boundaries accompanied by processes of granitization: on the levels of about 1 billion and 0.70–0.68 billion years. Less distinct is a boundary at 1.35 billion years given in

Table 10. Great and greatest boundaries of geological history

Stages in the development[a] of the Earth's crust and the greatest boundaries between them, million years	Boundaries within stages (great), million years
Stable geosynclinal-platform stage	200
570–600	400
Mobile geosynclinal-platform stage	680–700
1600–1800	1000
Unstable protogeosynclinal stage	2000–2200
2800–3000	2500–2600
Permobile stage	3500–3600
	4000

[a] The Earth's age about 4600 million years

the stratigraphic scale of the Precambrian period in the USSR (General problems... 1979). At the stable geosynclinal-platform stage, corresponding to the Phanerozoic, tectonic reconstructions and maximal regressions on platforms as well as orogenesis in Caledonian and Hercynian geosynclines, which are timed as about 400 and 270–200 million years ago, may be regarded as such boundaries.

Time intervals between the boundaries mentioned become shorter from the Archaean and the early Proterozoic to the Phanerozoic, from 400–500 to 180–200 million years ago. Yet, this shortening must not be regarded as absolutely proven, because there is a possibility that some boundaries are omitted as early stages in the development of the Earth's crust have been insufficiently studied. However, time intervals between the most important boundaries separating stages in the development of the Earth's crust seem also to diminish and this increases the probability of such a conclusion.

Within the Phanerozoic, the above-mentioned boundaries or reconstruction phases coincide with epochs of maximal regressions of the sea from continents. These regressions are clearly distinguished in plots of the change in areas occupied by the seas (Fig. 33, curves 1 and 2). They are also recorded in the change of level of the World Ocean, which is determined through statistical treatment of extensive seismological material (*Seismic Stratigraphy* 1977) (Fig. 34, 7).

Basic regressions correspond to epochs of orogenesis and granitization in geosynclines as well as to inversions of aulakogenes and acceleration of growth of structures on platforms. Transgressions correspond to epochs of quick sagging of geosynclines and also the laying of aulakogenes and manifestations of basic magmatism on platforms. At comparatively long time intervals, separating rapid development, transgressions, and regressions, there occurred an abatement of tectonic movements and a reduction in the growth rate of platform structures up to complete cessation.

Milanovsky (1979) believes that epochs of orogenesis, folding and inversions of aulakogenes, correspond to epochs of general compressions, and epochs of transgressions and sagging to phases of general expansion. It is necessary to add that, besides epochs of compression and expansion, there also exist epochs of abatement (rest) of tectonic movements. A regular alternation of epochs of com-

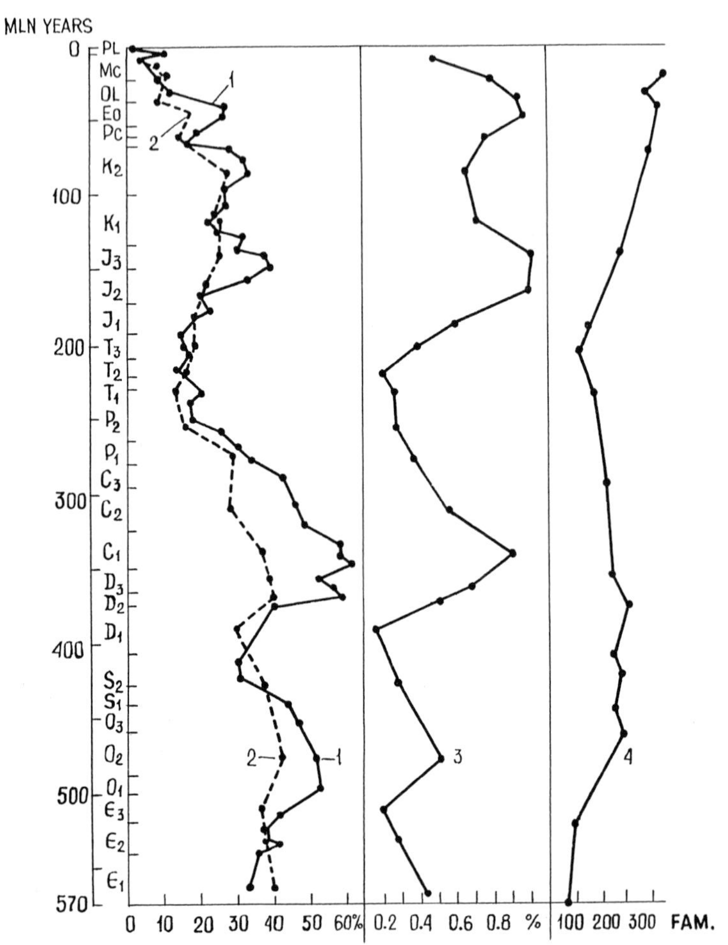

Fig. 33. Changes in areas occupied by seas for the USSR (*1*) and world continents (*2*), in contents of C_{org} in sedimentary rocks (*3*), and in the number of marine invertebrate families (*4*) in the Phanerozoic. *Abscissa* % of areas occupied by seas (*1*) and (*2*); % of C_{org} contents (*3*); numbers of families (*4*) (Data from Karogodin 1980 for *1*, Ronov 1976; Valentine 1973 for *2, 3*)

pression, expansion and rest forms large cycles corresponding to generally accepted cycles of tectogenesis (Fig. 34, 10).

Thus, transgressions and regressions are a sensitive quantitative index of the boundaries of geological history, well coordinated with a great number of tectonic events.

It is possible to single out systems of minor and more often alternating boundaries only in the Phanerozoic although, judging by the rhythmical structure of sections of Riphean deposits, they occurred also before. Generally speaking, with minor boundaries the systems become visible later. These minor boundaries are recorded on curves of changes in transgression areas, and even more vividly by a change in the level of the World Ocean (*Seismic Stratigraphy* 1977), for which an enormous amount of homogeneous and quantitative material is available (Fig.

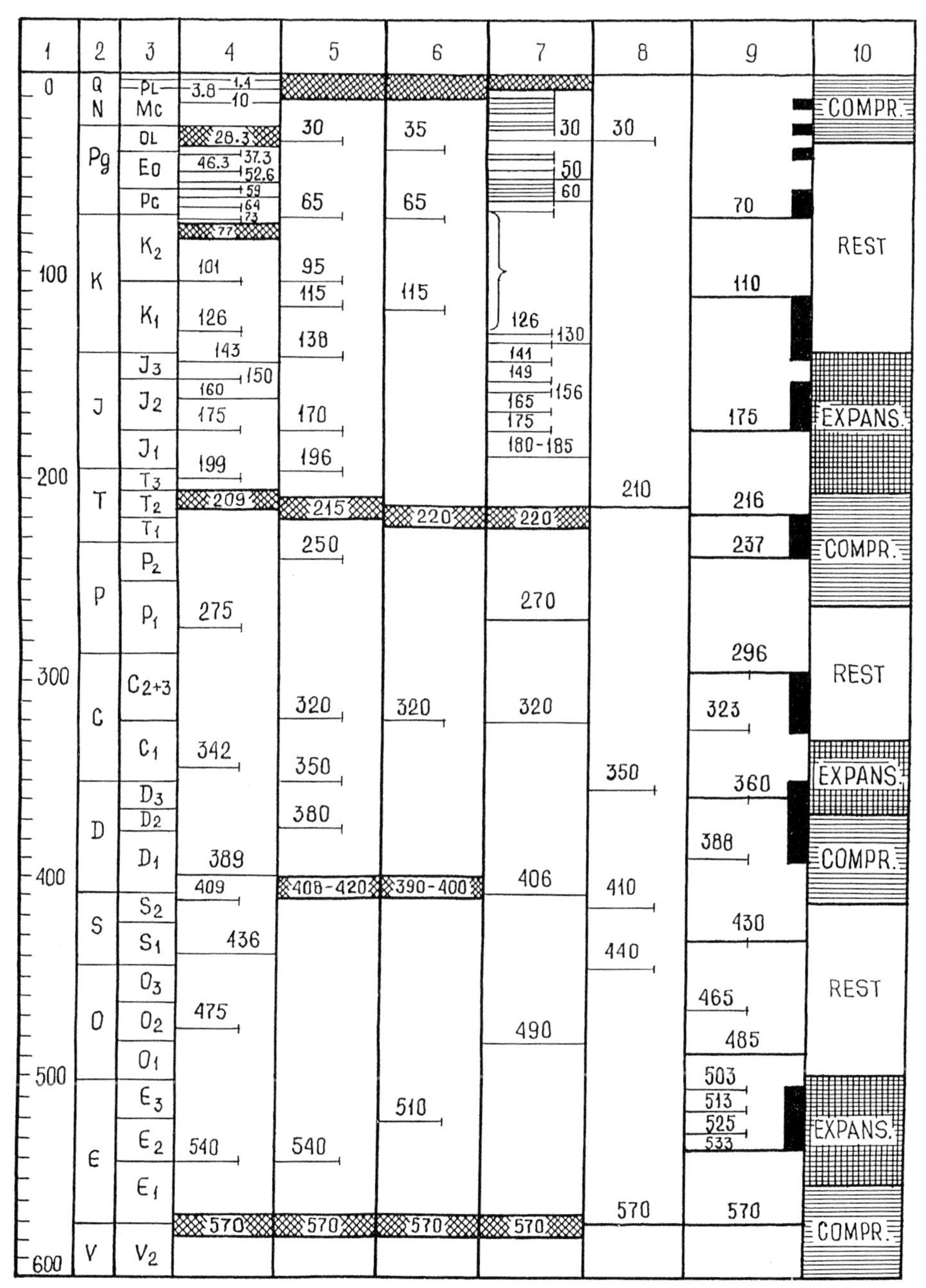

Fig. 34. Correlation of geological events in the Phanerozoic. *Vertical columns: 1* time in million years, stratigraphic scale; *2* systems; *3* branches; *4* calculated boundaries of the geological history; *5* main regressions for USSR territory (Karogodin 1980); *6* main regressions for continents of the world (Ronov 1976); *7* cycles of sea-level fluctuations (Data from *Seismic Stratigraphy* 1977); *8* change in the number of marine invertebrate families (Data from Valentine 1973); *9* paleomagnetic boundaries (Data from *Magnetostratigraphic Scale* 1976); *10* epochs of compression, expansion and rest (Data from Milanovsky 1979). *Shaded strips* in *columns 4–7* show largest boundaries; *black rectangles* in *column 9* indicate epochs of frequent change of polarity according to paleomagnetic data; *long horizontal lines* with numerals (time, million years) in *columns 4–8* are the largest boundaries (paleomagnetism hyperzones); *shorter horizontal lines* in the *same columns* are middle boundaries (paleomagnetism superzones)

34, 7). In the Paleozoic, stages of approximately 50–80 and 30–40 million years long are distinguished. In the Mesozoic, the duration of stages seem to shorten to 30–40 and 10–20 million years. In the Quaternary period, the shortest stages of dozens of thousand years long are noted. Boundaries between them are mainly drawn by using paleoclimatic evidence and data on alterations in fauna complexes connected with climatic changes.

Data on changes of the World Ocean level also show an increase in the frequency of boundaries of the geological history in its last stages. Stille (1964) concluded that, considering the early periods of tectonic development of our Earth, one can observe a strong concentration of orogenic manifestations and an increasing number of orogenic phases of decreasing mean intensity, up to the polyorogenic stage of recent geological time (p. 391).

In the development processes of systems, evolutionary and involutionary modes of development are distinguished (Vernadsky 1978). During evolutionary development in due course a decrease in boundary frequencies occurs from low hierarchical levels to high ones, and the involutionary mode is connected with an inverse tendency. For example, the evolutionary mode of development is growth from embryo to mature organism, and the involutionary mode is development without growth, e.g., the development of the Earth, with its mass remaining constant.

Thus, involutionary systems, the Earth included, should develop with an increase in boundary frequencies. As it has been already mentioned, the scanty data available on geological history confirm these conclusions.

Such an increase of the geological history rhythm, together with a loss of information on events of the distant past, make us compare boundaries of geological history with calculated critical boundaries, separately for the Precambrian, Mesozoic, Neogene, and Quaternary periods, considering correspondingly more and more minor boundaries.

Paleontological boundaries. Analysis of paleontological data is of special interest because it is paleontology that has given to science innumerable items of evidence of real significance of the evolutionary process. At the same time, paleontology rejected the concept of a gradual and smooth course in evolution and showed an unevenness of the evolution rates (e.g., Grant 1977).

Sokolov (1977) has pointed out the following most important events in the history of the development of the Earth's organic matter along the way to its Phanerozoic differentiation.

Appearance of the simplest eobiont systems on Earth, which were ready for self-reproduction under conditions of heterotrophic feeding (probably earlier than 4.25 billion years ago).

Appearance of photosynthesizing mechanisms in prokaryotic protobionts (probably already existing independently in parallel lines of evolution of bacteria and blue-green algae) that has opened the way to the biogenic accumulation of oxygen, primarily in the hydrosphere (3.8 billion years ago; blue-green algae from the "Ishua" sediments in Greenland).

Appearance of free oxygen in the atmosphere – immediately fixed by oxidizing reactions, probably after the glacial epochs of the Aphebian (2.2–2.0 billion years ago, Greenlandic microflora) – and its general increase, which seems to have prepared the rise of the most ancient eukaryots (1.9–1.6 billion years ago).

Appearance of reliable eukaryots that increased considerably their ability to form simple colonies and aggregations (1.6–1.35 billion years ago).

Transition from enzymatic metabolism (fermentation) to oxygen respiration: reaching of the Pasteur point (approximately 1% of free oxygen content in the modern biosphere) that exercised a revolutionary influence on the course of the evolutionary process and resulted in the beginning of mitosis and meiosis; appearance of the first Metaphyta, and pelagic and benthic Metazoa which are known only by the traces left by their vital functions in the late Riphean (1.0–0.9 billion years ago).

Wide spread occupation of the sea bottom and the pelagic zone by multicellular animals and plants after the late Precambrian (Lapland); glaciation on the border of the Riphean and Vend; strongly pronounced dominance of skeleton-free Metazoa represented, however, by very many main Phanerozoic types (period of $675 \pm 25 - 570$ million years ago).

"Population explosion" of marine skeleton-forming invertebrates – at the beginning of the classic Phanerozoic (Cambrian 570–500 million years ago).

Appearance of the protective ozone screen over the planet (about 10% of free oxygen content in the recent atmosphere). As far back as in the middle Ordovician, this created favorable conditions for the gradual occupation of land by vegetation and vertebrates, and created the possibility of their spreading from the Devonian to recent time.

In the Phanerozoic, the following basic periods of flora and fauna changes are distinguished (Rezanov 1980). In the late Permian period (230–250 million years ago) many groups of organisms died out, about 24 orders of fauna disappeared, and considerable impoverishment of the fauna took place during a further 10 million years.

Even more considerable change in flora and fauna was noted to have occurred in the late Mesozoic era, not only marine animals but also the dominating group of terrestrial quadrupeds, the dinosaurs, become extinct. It happened in the late Cretaceous, about 80 million years ago.

Let us consider now the quantitative data characterizing the development of the organic world. The materials available are limited. We have information on the dynamics of numbers of families, orders and classes of marine-shelf invertebrates (Valentine 1973). A curve of change in family numbers is analogous to curves of change in class and order numbers, though somewhat more detailed. A comparison of it with curves of transgressions and the content of organic carbon in rocks (see Fig. 33) permits us to speak of their general similarity. One can see a common maximum prevailing in the whole Paleozoic with a small decrease in the late Silurian and early Devonian; a steep minimum falling on the late Permian and Triassic; and at last, a new maximum associated with the Jurassic, Cretaceous and Paleogene. In addition, the curve of organic carbon content which, as it seems, should have been more related to the curve of the development of the organic world, occupies the intermediate position between the curve of transgressions decreasing in time and the growing curve of family numbers. As to the particular minima and maxima, it resembles more the curve of transgression. The latter can be explained by the fact that organic matter in continental sediments, whose amount increases in regression epochs, accumulates and is preserved less well than in sea sediments.

The fact that the appearance of new organismic forms is linked in time to the boundaries of geological history in the Precambrian, as well as to synchronous change in family numbers and ocean level, testifies to the general coincidence of boundaries in the development of living and nonliving matter.

Paleomagnetic boundaries. Paleomagnetic boundaries are cited from the Magnetostratigraphic Scale of the Phanerozoic (1976). In most cases, they are close to boundaries singled out in terms of geological data (Fig. 34, 9). For basic boundaries, the coincidence is quite close, but the less significant paleomagnetic boundaries mostly do not coincide with the geological ones. Epochs of frequent polarity change coincide also but poorly with geological events and, besides, are extended temporally. On the whole, the degree of correlation of paleomagnetic boundaries with both geological and calculated boundaries is lower than the correlation of geological with calculated boundaries.

The correlation of geological, paleontological and calculated boundaries: As it was shown above, the development unit selected by calculation is limited by two basic boundaries which have the greatest power of manifestation. It includes one intermediate boundary with somewhat less power, five boundaries with secondary power of manifestation and, lastly, a boundary of the development rate in the reconstruction period at its transition from the previous link to the next one. On this boundary no reconstructions occur and that is why this cannot be regarded as a boundary comparable in its nature to the rest of the boundaries considered. The sequence of all these boundaries is given in Fig. 34.

To find the main and other boundaries within the geochronological scale, as given in Table 4, we should know the duration of one of the even cycles T^K. This can be used to determine both the position of consecutive boundaries, according to Eq. (81), and the duration of even cycles of various orders according to Eq. (82).

We can consider two approaches to the estimation of the value T^K:

1) The position of some well-studied geological boundary should be taken as an initial one and used to find T^K according to Eq. (81).
2) The duration of some well-studied even cycle T^K, determining the critical boundaries of the order to be investigated, should be taken as an initial value.

If, in accordance with the first approach, the upper border of the Precambrian (570–600 million years ago) is taken for the initial critical boundary, we obtain the duration of the cycle from the formula (81):

$$T^k = \frac{t^k}{3.16} = \frac{570}{3.16} = 181 \text{ million years} .$$

If the termination of the Hercynian cycle of tectonic development, falling after many authors in the middle Triassic, is taken for the initial boundary, we obtain

$$T^k = \frac{210}{3.16} e = 180.6 \text{ million years} .$$

In accordance with the second approach, the duration of large cycles of transgression (V. Nalivkin 1975), equal to 180–200 million years, is taken as the initial duration of a T^k cycle.

Thus, the estimates obtained by different approaches give similar values.

To estimate the position of critical boundaries within the geochronological scale, we take the duration of the even cycle as

$$T^k = 180 \text{ million years} .$$

Using the formula (81), we determine the position of the main boundary, coinciding with this cycle:

$$t^k = 3.16 \cdot 180 = 569 \text{ million years} .$$

Positions of the preceding and the following main irregular critical boundaries, calculated according to Eq. (70), will be given by the sequence: 28, 77, 209, 569, 1548, and 4210 million years. According to Eq. (82), the given sequence of the basic irregular critical boundaries corresponds to a sequence of even cycles T^k: 9, 24, 66, 180, 490 and 1330 million years. Then, the even boundaries of the order K within each irregular range are described by the expression

$$t^k - T^k = 2.16 T^k ,$$

and, accordingly, the consecutive intermediate boundaries will be 53, 143, 379, 1058, and 2878 million years. A collation of the main irregular critical boundaries and even boundaries of the order K with the principal and the main stages in the development of the Earth is given in Fig. 1.

Hence, we have analytically determined an elementary link ("unit of development") in the geological process, where synchronization of the main even boundaries of the orders K and K–1 with irregular alternating boundaries occur. Within this unit, we have also determined the position of the time of termination of a reconstruction phase of the preceding unit. A scheme of the structure of this unit is given in Table 4. We have further determined a relative power of manifestation of different boundaries within the unit of development. On repetition, such standard units compose a chain, corresponding to the process of development of the system in question as a whole. In this chain, every following link is e times shorter than the preceding one. And at least, we have collated the limits of one of such links and the geochronological scale. At this collation, the basic critical boundaries of the irregular sequence coincided with the main stages in the Earth's development. The system of elementary calculated links was associated with the boundary at 570 million years, that is, with the lower boundary of the Phanerozoic. Here, the duration of the basic even cycle of development is 180 million years (duration of a galactic year). The duration of the units of development diminishes. In the Precambrian, two such units are distinguished.

Accuracy in the calculation of boundaries depends on the accuracy of the initial boundary by which an attachment to the system of geological boundaries is made, or the correctness of evaluation of the basic even cycle duration in geological history. The accuracy of the calculated system of boundaries can be examined by a degree of coincidence of calculated boundaries with geological boundaries within the entire geological history, that is, during 4–5 billion years. As it will be shown below, such a coincidence is really observed and, therefore, the initial values used for attachment – a boundary at 570 million years and a cycle duration of 180 million years – were chosen right and the construction of the calculated system of boundaries appears to correspond to the natural processes.

The time of manifestations of boundaries on early stages of the Earth's development is evaluated with a low accuracy. This depends on the extension of geological processes occurring on boundaries, on indistinct criteria of their evalution, the temporal shift of the same developmental stages in different regions of the world, as well as on a great absolute error in determining absolute time. At later developmental stages, the error decreases sharply. Apparently, the error for all stages will be no less than 5%.

Accuracy in the determination of paleontological boundaries, which are also not momentary, depends on the rate of changes in fauna and on whether this change occurs simultaneously in all regions of the globe. In the Precambrian, these rates were low. Thus, the duration of existence of coeval communities of calciphyte algae was considerably longer in the Precambrian than in the Mesozoic and Cainozoic.

A collation of the Precambrian boundaries is given in Table 11. In column 2, calculated boundaries are given, the boundary at 570 million years being used as an initial one for the calculated system of boundaries. Column 3 shows a position of geological boundaries according to the sources cited in the next columns, and some additional data. Column 3 is the initial one for comparison with the calculated boundaries. In column 4, boundaries adopted by the international stratigraphical scale for the Precambrian (Sims 1979) are given. It should be noted that almost all these boundaries are somewhat rejuvenated compared with the stratigraphical scale of the Precambrian in the USSR (General problems... 1979) given in column 5. Column 6 cites boundaries adopted in the volume *The Russian Platform* of the monograph *The Geological Structure of the Territory of the USSR and Regularities in the Location of Deposits* (1968). In column 7, main epochs of granitization are shown (after Krats et al. 1980). And in column 8, the main levels of change in the most ancient fauna and flora are given, after Sokolov (1977).

It follows from Table 11 that the results obtained for major geological boundaries determined in different regions and by different indications are close to each other. Age fluctuations seldom exceed the accuracy of their determination. This allowed us to compose the general scheme given in column 3. A comparison of the data from the scheme with the calculated boundaries (column 2) shows their good convergence (Table 12).

In the comparison, calculated boundaries at 2872 and 3017 million years are accepted as coincident with geological boundaries at 2.9 and 3.0 billion years, because a divergence between them is close to the accuracy of age determination (±100 million years). A certain disagreement between calculated and geological boundaries observed in the upper part of the scale is, most probably, connected

Table 11. Boundaries in the Precambrian (age in million years). Greatest geological boundaries are *boldface*

Strati-graphical scale	Calculated boundaries		Geological boundaries					
	Boundary rank and age		Accepted boundaries	International stratigraphical scale	Stratigraphical scale of the USSR	*The geol. structure of the USSR. The Russian platform* 1968	Epochs of granitization after Kratz et al., 1980	Appearance of various groups of organisms by Sokolov, 1977
Vend	Basic (initial)	570	**570 (±20)**	570	570	570		570–675
			680–700 (±20)		680–700	680		
	Secondary	748						
	Secondary	929		900				
Upper	Intermediate	1056	1000 (±50)		1050		about 1000	900–1000
	Secondary	1109						
Proterozoic	Reconstructional	1187						
	Secondary	1291	1350 (±50)		1350			
	Secondary	1467						1350–1600
(Riphean)	Basic	1546	**1600–1800 (+50)**	1600	1650	1650–1800	1700–1800	1600–1900
Lower and	Secondary	2034	1900–2100 (±100)	2100	1900		1900–2100	2000–2200
middle					2300	2200		
Proterozoic	Secondary	2526	2500–2600 (±100)	2500	2600	2600	2500–2700	
	Intermediate	2872	**2900–3000 (+100)**	2900	3000	3000	3000–3200	
	Secondary	3017						
Archean	Reconstructional	3225						
	Secondary	3510	3500–3600	3500	3500		3600–3800	3800
	Secondary	3990	**4000**			4000		
Katarchean	Basic	4200	4600[a]					

[a] Age of the Earth

Table 12. Collation of the Precambrian boundaries. In *brackets*, percentage of coincident boundaries is given

Coincidence characteristics	Coincidence of accepted boundaries with calculated estimates		Coincidence of calculated estimates with accepted boundaries	
	All boundaries	Greatest boundaries	All boundaries	Basic and intermediate boundaries
Total number of boundaries	10	4	17	5
Coincident	9 (90)	4 (100)	12 (70)	5 (100)
Noncoincident	1–2	0	5	0
Coindicent in rank	7 (77)	3 (75)	(75–83)	3 (60)
Noncoincident in rank	2	1	2	2

with a gradual shortening of intervals between calculated boundaries and the elimination of the power of their manifestation. There was a tendency to establish geological boundaries of the same power of manifestation, therefore they were arranged with more regular time intervals. Reconstructional boundaries differ from other boundaries in their character: they reveal only the time of reconstruction accomplishment at transition between successive periods. Reconstructional boundaries have no analogs among geological boundaries and are not included in the number of divergencies.

A comparison of Phanerozoic boundaries (without the Neogene and Quaternary) is given in Table 13. In column 4, the time of sea regressions on USSR territory is given, calculated by Karogodin (1975). These regressions are evaluated in more detail for the USSR than for other regions of the world (Fig. 34). Their reliability is confirmed by a good coincidence with world regressions and fluctuations in the level of the World Ocean (*Seismic Stratigraphy* 1977). In reality, maximal values of regressions do not fall on the middle parts of stratigraphical intervals, as it is shown in Fig. 33, but on the lower boundaries of these intervals, and the range itself of the sea-level fluctuations is considerably greater than that shown on the given curves, which were plotted according to average data for time intervals of 7–8 million years. It should be borne in mind that regression values for the Oligocene are given schematically although in reality they are more differentiated.

In column 5, moments of minimal levels of the World Ocean are given (*Seismic Stratigraphy* 1977, pp 84–85). For the Mesozoic and Paleogene, global "supercycles" of the ocean level fluctuations and "cycles" of a low rank are given. For the Paleozoic, only "supercycles" are shown; in Table 13 they are underlined. The supercycles are also taken as basic in comparison with calculated boundaries. The above mentioned work does not give "cycles" for the Cretaceous period.

Column 3 generalized the estimates, it includes the most reliable boundaries determined according to summarized data (columns 4–6) or, more rarely, gives very distinct boundaries singled out by one feature. Limits of errors of radiolog-

Table 13. Boundaries in the Phanerozoic (without Neogene and Quaternary; age in million years). The main geological boundaries are *boldface*

Stratigraphical scale			Calculated boundaries		Geological boundaries			
			Boundary rank and age		Accepted boundaries	Maximal regressions for the USSR	Regressions from *Seismic Stratigraphy* (1977)	Change in family numbers
			Basic	28.3	30 (± 1)	30	**30**	30
O1 + N + Q		37	Secondary	37.3			37	
							40	
	Eoc		Secondary	46.3			45	
			Intermediate	52.6	50 (± 3)		**50**	
							52	
Pg		53	Secondary	55			53.5	
							56	
							59	
	Pc		Reconstructional	59			**60**	Change of fauna
			Secondary	64	60 – 65 (± 3)	65	65	
		67	Secondary	73			Gap	
	K_2		Basic	77				
K		100	Secondary	101	95 – 100 (± 4)	95 (?)	**98**	
	K_1		Secondary	126		115	126	
		137					**130**	
			Intermediate	143	137 (± 5)	135 – 138	141	
	J_3		Secondary	150			150	
		161	Reconstructional	160			165	
J	J_{1+2}	170	Secondary	175	175 (± 5)	170	175	
							190	
		195	Secondary	199		196	198	
	T_3							
T	T_2		Basic	209	**200 – 220 (± 10)**	**215**	**220**	**210**
	T_1							
		230						
	P_2		Secondary	275	265		**270**	
P	P_1							
		285				320	**320 – 325**	
	C_{2+3}							
C	C_1		Secondary	342	325 (± 10)	350		350?
		350			350 (± 10)			
	D_3	385	Intermediate	389	380 (± 10)	380		
D	D_2							
	D_1							
		406	Secondary	409	**400 – 420 (± 10)**	**408 – 420**	**406**	410
	S_2		Reconstructional	436				440
S	S_1							
		440						
	O_3		Secondary	475	480 (± 15)		490	Occupation of land
O								
	O_2	480						
	O_1	510						
	$ε_3$	545	Secondary	540	**540 (± 15)**	520 – 540		
ε	$ε_2$							
	$ε_1$	570	Basic (initial)	570	570 (± 20)	**570**	**570**	**570**

Table 14. Collation of the Phanerozoic boundaries. In brackets, percentage of coincident boundaries is given

Coincidence characteristics	Coincidence of accepted boundaries with calculated estimates		Coincidence of regressions from *Seismic Stratigraphy* (1977) with calculated estimates		Coincidence of calculated estimates with accepted boundaries		Coincidence of calculated estimates with regressions from *Seismic Stratigraphy* (1977)	
	All boudaries	Main boundaries	All regressions	Regressions of "supercycles"	All boundaries	Basic and intermediate boundaries	All boundaries	Basic and intermediate boundaries
Total number of boundaries	15	4	24	12	20	6	20	6
Coincident	14 (93)	4 (100)	17 (71)	10 (83)	14 (70)	6 (100)	17 (85)	5 (83)
Noncoincident	1	0	7	2	6	0	3	1
Coincident in rank	10 (71)	4 (100)	10 (57)	4 (40)	10 (71)	3 (50)	10 (59)	4 (80)
Noncoincident in rank	4	0	7	6	4	3	7	1

Table 15. Boundaries in the Neogene (age in million years). Basic geological boundaries are *boldface*

Stratigraphical scale (*Seismic Stratigraphy* 1977)			Calculated boundaries		Geological boundaries				
			Boundary rank and age		Regressions (*Seismic Stratigraphy* 1977)	Stratigraphical boundaries (Berggren and Convering 1973)		Regressions (Krasnov 1977)	Cooling (Krasnov 1977)
1			2		3	4		5	6
Quaternary	Eopleistocene	Calabrian	Basic	1.41		Calabrian			
			Secondary	1.85	1.8	______	1.8	1.7 – 1.8	1.6 – 1.8
			Secondary	2.3					2.3 – 2.5
			Intermediate	2.62				2.5	2.6 – 2.8
			Secondary	2.73	**2.8**	Piacenzian			

Neogene	Pliocene	Piacenzian	Reconstructional	2.92					
			Secondary	3.2			3.2		
			Secondary	3.64					3.5 – 3.6
			Basic	3.83	3.8	Zanclian			
		Tabianian	Secondary	5.04	5.3		5	5 – 6	
	Miocene	Messinian	Secondary	6.26	6	Messinian			
			Intermediate	7.12	**6.6**		6.6		
		Tortonian	Secondary	7.5					
			Reconstructional	8					
						Tortonian			
			Secondary	8.7					
			Secondary	9.9					
			Basic	10.4	11		11		
		Serra-vallian	Secondary	13.7	13	Serravallian			
							13 – 14		
		Langhian			15.7	Langhian			
			Secondary	17					
			Intermediate	19.3		Burdigalian			
		Bur-digalian	Secondary	20.3	19		19		
			Reconstructional	21.7		Aquitanian			
			Secondary	23.6					
		Aquitanian			24		22.5		
			Secondary	26.9					
					26	Chattian			
Paleogene	Oligocene	Chattian	Basic	28.3					
					30				

ical age determination are given in brackets. Great boundaries mentioned at the beginning of the present section are underlined. The boundary at 30 million years, which can be associated with the beginning of the Alpine folding, is also classified as one of the great boundaries.

A comparison of boundaries accepted for the Phanerozoic and "supercycles" of the ocean level fluctuations with calculated boundaries shows approximately the same convergence as for the Precambrian. The coincidence is as high as 70%–80% and even 100% for the main boundaries (Table 14). As for the Precambrian, the boundaries, with a divergence which was less than, or slightly exceeded, the error in radiometric age determination, were regarded as coincident. Reconstructional boundaries were not taken into account at the collation.

The most obvious disagreements between calculated and geological boundaries are, on the one hand, the geological boundary at 320–380 million years between the early and the middle Carboniferous periods, which has no analog among calculated boundaries and, on the other hand, the main calculated boundary at 77 million years (within the late Cretaceous), which is not distinctly reflected in geological history. The reasons for these divergences are not clear. Moreover, as in the Precambrian, in the upper part of the scale, calculated boundaries are more frequent than geological ones, though cycles of lower rank in the ocean level, determined according to seismological data, occur here no less frequently.

There is a good coincidence between boundaries of sections and systems of the stratigraphic scale and calculated boundaries. The coincidence is about 70%. Major discrepancies fall on boundaries where no important events occurred; T and P, D_3 and D_2, S_2 and S_1, O_3 and O_2, C_3 and C_2. As mentioned above, boundaries at 320–330 and 77 million years manifest the main divergencies.

A collation of Neogene boundaries is given in Table 15. Unfortunately, we failed to find this time complete data on ages of the main regressions and on climatic changes (column 5 and 6, Table 15). Evidence available for the late Neogene is well associated with the position of boundaries between stratigraphical units. There are more complete data on level fluctuations in the World Ocean (*Seismic Stratigraphy* 1977) and on boundary age in the stratigraphical scale.

It should be noted that maxima of regressions in the World Ocean level, determined by seismological data, almost perfectly coincide with the stratigraphical boundaries accepted in this work. Additional regressions on the levels of 13 and 19 million years coincide with the stratigraphical boundaries accepted by Berggren and Convering (1973). Only a regression on the level of 3.8 million years is not reflected in the stratigraphical scales. The stratigraphical scale given here differs from the scale in *Seismic Stratigraphy* (1977) in the position of boundaries of the Calabrian, Serravallian and Aquitanian. It is interesting that these boundaries are shifted exactly by one regression. Therefore, we can believe that one and the same natural phenomenon was interpreted differently by different authors. Since maxima of regressions coincide well with the stratigraphical scale, we will give a comparison of calculated boundaries with the maxima of regressions only (Table 16).

As for the above mentioned units of geological history, main units coincide best of all, all other units coincide somewhat worse, and the coincidence of

Table 16. Collation of Neogene boundaries. In *brackets,* percentage of coincident boundaries is given

Coincidence characteristics	Coincidence of regressions from *Seismic Stratigraphy* (1977) with calculated estimates		Coincidence of calculated estimates with regressions from *Seismic Stratigraphy* (1977)	
	All cycles	Supercycles	All boundaries	Basic and intermediate boundaries
Total numbers of boundaries	14	3	22	7
Coincident	13 (93)	3 (100)	13 (59)	6 (85)
Noncoincident	1	0	9	1
Coincident in rank	10 (77)	3 (100)	9 (70)	3 (50)
Not coincident in rank	3	0	4	3

boundary ranks is appreciably worse. The boundary between the Tertiary and Quaternary periods should be noted among the main divergencies: according to calculations it must be on the level of 1.41 million years but geological indications show it on the level of 1.7–1.9 million years. The position of this boundary is also a matter of controversy among stratigraphers.

Boundaries of the Quaternary period are given in Table 17. Since the Quaternary period has five units of development and many minor boundaries which are not distinguished in the geological history considered before, only basic and intermediate calculated boundaries were collated, secondary boundaries excluded.

Stratigraphy of the Quaternary period is mainly climatostratigraphy. Its units correspond mostly to periods of warming and cooling. The cooling periods correlate well with regressions of the World Ocean selected by Fairbridge (from Klige 1980) for the upper Pleistocene and Holocene. Such coincidence of stratigraphical boundaries, climatic units and regressions (Table 18) makes it possible to restrict our analysis to the collation of calculated boundaries only with stratigraphical boundaries (column 3). A superposition of stratigraphical boundaries with regressions allows us to use a single approach to establishing geological boundaries within the entire Phanerozoic. Stratigraphical boundaries and their collation with the chronological scale are given according to Krasnov (1977).

Coincidence of calculated and geological boundaries for the Quaternary period is appreciably worse than for the above considered time intervals (Table 18). Most discrepancies fall on the interval from 0.9 to 1.8 million years, i.e., on the early Quaternary period.

The collation of boundaries obtained through calculation with boundaries accepted in geological history and established by different characteristics has shown a good coincidence.

Better coincidence (about 85% on the average) is observed for the most strongly pronounced boundaries. Coincidence for all boundaries is somewhat worse, about 75%. The worst coincidence is found in comparison by ranks, when main geological boundaries coincide with secondary calculated boundaries and vice versa. Coincidence in ranks averages 60%–65%.

A strict determination of the coincidence degree is pointless because, first, a statistical sample is inadequate for it and, second, geological boundaries are extended in time and determined often indistinctly.

Table 17. Boundaries in the Quaternary period (age in thousand years). Basic geological boundaries are *boldface*

Stratigraphical scale		Calculated boundaries		Geological boundaries		
		Boundary rank and age		Age of stratigraphical boundaries (Krasnov 1977)	Maximal regressions after Fairbridge (from Klige 1980)	Cooling (Krasnov 1977)
Holocene		Basic	9.5	10	10–15	
		Intermediate	17.6			
Upper section (Pleistocene)	Upper part	Basic	25.8	23	20	12–24
		Intermediate	48	50	50	47–55
		Basic	70.2	**70**	60–70	59–64
				100	115	115–125
		Intermediate	130	130		
		Basic	191	180	182	180–230
				240	233	
	Middle part	Intermediate	354	380	380	
		Basic	518	**500**		390–500
	Lower part	Intermediate	960	**900**		800–900
Lower section (Eopleistocene)	Upper part	Basic	1410	**1250**		1150–1250
	Lower part			**1550**		1550–1800
				1800		

A tentative estimation of the coincidence degree shows that boundaries coincide on an average in 75% cases. Such a high degree of coincide cannot be accidental. Therefore, analytical calculation reflects real correlations in the arrangement of revolutionary events in geological history.

It is necessary to mention that one cannot expect to find an absolute coincidence between calculated and geological boundaries. Apart from errors in determining the position of geological boundaries, the latter are just unstable. They depend on a great number of various conditions which influence the time of their appearance. Hence, misrepresentations of the general objective regularity are inevitable. This regularity shows itself only as a tendency but not as a strict rule.

A system of calculated boundaries reflects the general objective regularity in the development of natural objects but does not take into account the influence which the development of some objects exerts on others. This influence can work in two ways: first, revolutionary reconstructions in the development of some objects can, like a starting device, accelerate or decelerate the beginning of the same reconstruction of other objects, and second, such reconstructions of some objects can cause a change in the rate and character of development of others, in other

Table 18. Collation of boundaries of the Quaternary period. In *brackets*, percentage of coincident boundaries is given

Coincidence characteristics	Coincidence of stratigraphical boundaries (Krasnov 1977) with calculated ones		Coincidence of calculated boundaries with stratigraphical ones
	All stratigraphical boundaries	Basic stratigraphical boundaries	Basic and intermediate boundaries
Total number of boundaries	14	6	11
Coincident	9 (64)	3 (50)	8 (73)
Noncoincident	5	3	3
Coincident in rank	4 (50)	2 (100)	4 (50)
Noncoincident in rank	4	0	4

words, it can result in the unevenness of processes. In the first case, a misrepresentation of the objective regularity of development will happen. Certain divergencies of calculated and actual geological boundaries mentioned above result, probably, from such a misrepresentation. In the second case, there will be most likely no divergencies and disturbances of objective regularities of development, because the changes in both objects will be synchronized and become a common system.

Thus, a regular connection between the basic characteristics of a developing system and the positions of its critical boundaries can be used not only for living objects, but also for nonliving ones, for example the Earth's crust.

A particular conclusion for geology is: in choosing a position for stratigraphical boundaries and dating tectonic events, the calculated boundaries and their arrangement by force should be taken into account. These boundaries, however, may serve only for reference in the analysis of factual material and in the determination of the real time of reconstructions. Besides, a knowledge of objective regularities of boundary alternations is very important for dating still undated geological events, as well as for various classifications and schemes.

Lastly, it is very important that the main deposits of minerals are connected with rocks formed on the boundaries. This is observed both in the Precambrian geosynclinal formations and in the Phanerozoic platform formations. Most deposits of oil, gas, coal, bauxites, phosphorites and other minerals are connected with epochs of intervals which coincide with the boundaries of different scales in geological history.

4.2 Temporal Rhythmicity of the Solar System

To determine the characteristics of a developing system, in conformity with the unit of elementary link of development, we must know the duration of one of the basic even rhythms playing the role of a tact mechanism, a "metronome of devel-

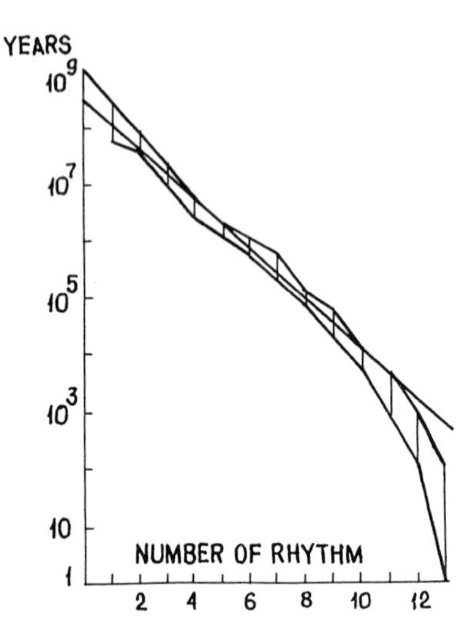

Fig. 35. Duration of rhythms of geological processes (Data from Krasnov 1977) in collation with the geometric progression with module e (*straight line*). *Abscissa* ordinal numbers of rhythms; *Ordinate* rhythm duration (logarithmic scale); *vertical straight lines* determine the limits of change of rhythm durations, according to stratigraphical data

opment." The galactic year can serve as such a standard, knowledge of its value enabled us to calculate the boundaries of the geochronological scale (Zhirmunsky and Kuzmin 1982). This showed that large-scale transformations in the processes of development of the Earth and the biosphere are determined by a system of hierarchically interrelated rhythms, and knowledge of one of them makes it possible to identify others. Rhythm durations singled out in the Earth's geological history in function of their rank number are presented (in a logarithmic scale) in Fig. 35 according to Krasnov (1977). The solid line denotes here a geometric progression with module e calculated from the duration of the galactic year in accordance with Eq. (3). The figure shows that in a range of rhythm durations from 100 millions to 10000 years a correlation between the average rhythm durations in geological processes corresponds to a geometric progression with the module e. The range of durations below 10000 years is inadequately expressed in geological data, and, consequently, a sequence introduced here would evidently have a different precision level.

Let us consider the rhythms determining the distribution of the planets around the Sun in the solar system. The Sun, as a central body, will be regarded as directing the basic uniform rhythm determined by a period of rotation around its axis on the equator $T^K = 25.36$ days (Childs 1962). Then the basic synchronized boundary of the unit of development will be described by Eq. (2).

$$t^K = \frac{2e}{e-I} T^K = 3.16\, T^K = 3.16 \cdot 25.36 = 80.1 \text{ days} \ ,$$

which is close to the period of Mercury's revolution around the Sun. Forming a further sequence of basic critical boundaries and reconstruction phases of the unit of development (see Table 7), we obtain a sequence of values given in Table 19. These values indicate that the revolution periods of the planets in the solar system (Mercury, Venus, Mars, Asteroids, Jupiter, Saturn, Uranus, Pluto; Fig. 36) are close to a geometric progression with the modulus e (the mean real value is

Table 19. Comparison of actual and calculated periods of revolution of planets of the solar system around the Sun

K	Planet	Revolution period t^K, days		Relative error, %	$\ln t^K_{act}$	$\ln t^K_{act}/t^{K-1}_{act}$
		t^K_{act}	$t^K_{cal} = e\ t^{K-1}$			
0	Mercury	88	80.1 [a]	−9	4.48	
1	Venus	224.7	217.8	−3.1	5.42	0.94
	Earth	365.3				
2	Mars	687	592	−13.8	6.53	1.11
3	Asteroids	1633 [b]	1609	−1.5	7.40	0.87
4	Jupiter	4332	4375	1	8.40	1.00
5	Saturn	10759	11893	10.5	9.28	0.88
6	Uranus	30685	32330	5.4	10.33	1.05
	Neptune	60189				
7	Pluto	90465	87882	−2.9	11.41	1.08
						$m^c = 0.99$

[a] The initial value
[b] The mean geometric value for the asteroid belt
[c] Mean value $\ln (t^K_{act}/t^{K-1}_{act})$

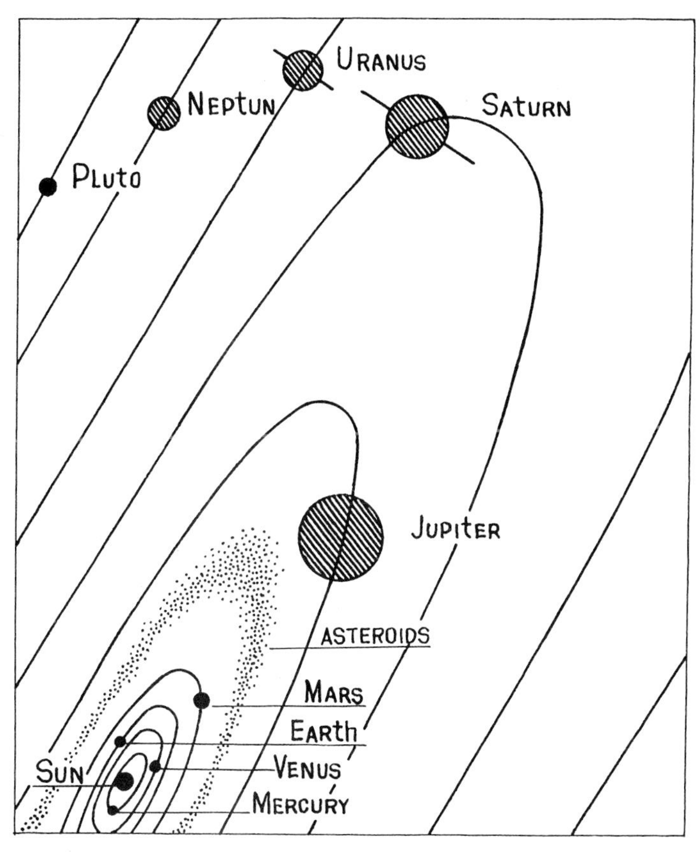

Fig. 36. Scheme of the solar system (Data from Kaufman 1982)

2.69). The Earth and Neptune drop out of this series, which is indicative of their particular position in comparison with other planets during the formation of the solar system and in our days.

Basic boundaries, in accordance with the correlation (3), form a geometric progression with the modulus e. Then, the position of the first synchronized boundary (80.1 days) being known, the next basic boundaries can be defined from the dependence:

$$t^{K} = e t^{K-1} .$$

The range of the unit of development from e^{e} to e^{3} is a reconstruction phase. It is here that a qualitative change of the system characteristics occurs. In fact, in a range between e^{e} and e^{3}, with respect to the revolution period of Mercury, a planet nearest to the Sun, there is a belt of asteroids. The suggestion was made that the planet Phaeton had existed there, which broke to pieces later on. Disintegration of planets in the zone of reconstruction is quite probable. This zone can include asteroids even along with planets. This is confirmed by the fact that a new unit of development that begins at the end of the asteroid belt has a reconstruction zone located before Uranus. It is in this very range that Chiron, a small planet of an asteroid type, has been recently discovered (*Astronomicheskii Ezhegodnik* 1980, p. 184).

4.3 Synchronization Rhythms of Motions of the Earth and Other Solar System Bodies

It follows from the previous section that the Earth has a particular place in the solar system. It has been found that the Earth does not fit into the geometric progression formed with the modulus e by periods of revolution of the planets of the solar system around the Sun.

The question arises how do characteristics of the revolution periods of the Earth and other solar system bodies relate to each other. Synchronization of the motion of the Sun, the Moon, and the Earth was known even in ancient times and used as a basis of calendar systems (Seleshnikov 1977).

Thus, cycles of the lunar calendar imply the synchronization of days and the lunar astronomical year, including 12 synodic lunar months lasting 29.53059 days each, i.e., the lunar astronomical year has a period of

$$T = 29.53059 \text{ days} \times 12 = 354.36708 \text{ days} .$$

Synchronization is achieved through approximating the period by a series of common fractions (Khinchin 1961). The fraction 0.36708 was expressed as a series of approximating common fractions (Seleshnikov 1977): $\frac{1}{2}, \frac{1}{3}, \frac{3}{8}, \frac{4}{11}, \frac{7}{19}, \frac{32}{87}, \ldots$.

The use of approximating fractions makes it possible to find such integral numbers, whose relation is most proximal to the original decimal fraction at a given accuracy level. The approximating common fraction $\frac{3}{8}$ determines an eight-year "Turkish" cycle (8 lunar years include 3 leap years with 13 lunar

Table 20. Convergent fractions of lunar and solar-lunar calendars and 1/e

Calendar	Convergent fractions						
Lunar	1/2	1/3	3/8	4/11	7/19	11/30	29/70
Solar-lunar	1/2	1/3	3/8	4/11	7/19	123/334	376/1021
1/e	1/2	1/3	3/8	4/11	7/19	32/87	39/106

months, instead of 12) and the approximating common fraction $\frac{11}{30}$ defines a thirty-year "Arabian" cycle (30 lunar years comprise 11 leap years).

Solar-lunar calendars have one basic cycle that coordinates the tropical year (365.24220 days) with a synodic month (29.53059 days). Dividing the duration of the tropical year by the synodic month,

$$\frac{365.24220}{29.53059} = 12.368266\ldots$$

we obtain 12 whole lunar months and a fraction to be compensated by leap years. To define leap years, the fraction 0.368266 should be expressed by approximating common fractions $\frac{1}{2}, \frac{1}{3}, \frac{3}{8}, \frac{4}{11}, \frac{7}{19}, \frac{123}{334}, \ldots$.

The majority of solar-lunar calendars are based on a cycle defined by the approximating common fraction $\frac{7}{19}$. In this cycle, termed the metonic cycle, 19 solar years are equal to 235 lunar months.

It is worth noting that the approximating common fractions which can determine cycles in lunar and solar-lunar calendars differ but insignificantly from the value 1/e, where e is the Napierian number. As it follows from Table 20, the first five approximating common fractions used as a basis for lunar and solar-lunar calendars are the same as those for the value 1/e. This follows from the numbers that underlie the cycles of lunar and solar-lunar calendars: 0.36706 for lunar calendars, 0.368266 for solar-lunar calendars, and 0.367879 = 1/e.

Taking 19 years for the basic critical period (since later on the approximating common fractions for the lunar and solar-lunar calendars diverge), we estimate the onset of the reconstruction phase of the unit of development with

$$19 \cdot e^e : e^3 = 14.3 \text{ years} .$$

Then the previous reconstruction phases will be represented by ranges given in Fig. 37.

In the Maya calendar (Seleshnikov 1977), which included 20 months of 13 days each, 260 days correspond to a short year. As seen from Fig. 37, 13 days cor-

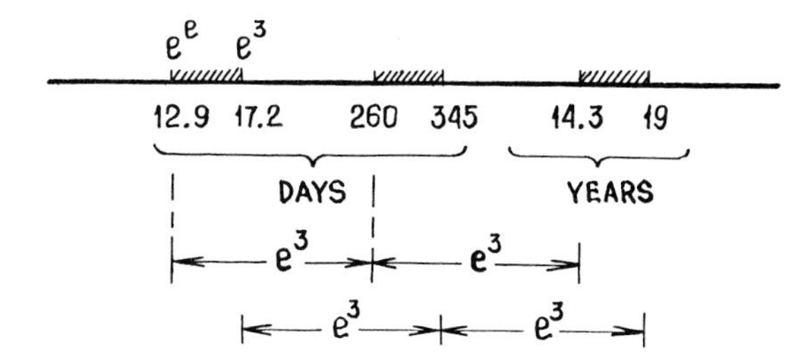

Fig. 37. Scheme of calculation of the reconstruction phase with the use of the age of synchronized boundary t^K = 19 years; *Shaded ranges* denote reconstruction phases

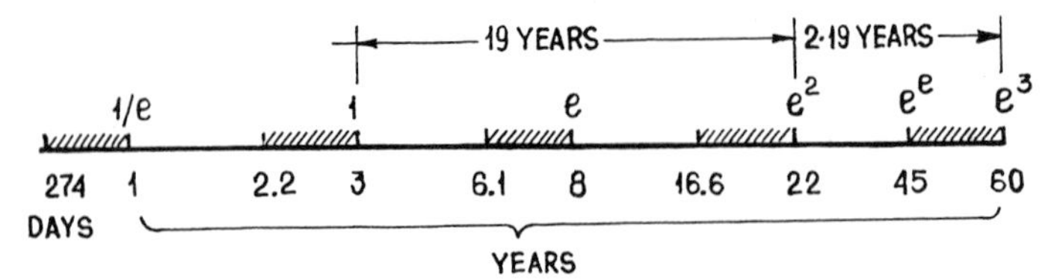

Fig. 38. Scheme of formation of basic boundaries and reconstruction phases of the unit of development with the use of duration of the boundary $T^K = 19$ years

respond to the duration of the starting period of the reconstruction phase, which differs from 260 days by a unit of development $e^3 = 20.085$.

Thus, the cycles of solar, lunar and solar-lunar calendars are obtained on synchronization of the periods of the Earth's rotation, the Earth's revolution around the Sun, and the Moon's revolution around the Earth. Cycles of the Maya calendar are obtained from the same initial data, with the structure of the elementary link of development taken into consideration.

Let us analyze a 19-year cycle as the duration of a regular rhythm, giving a sequence of critical boundaries in accordance with the unit of development ($T^K = 19$ years). Then from condition (81)

$$t^K = 3.16\, T^K = 3.16 \cdot 19 \text{ years} = 60 \text{ years} \ .$$

Hence, the unit of development assumes the appearance presented in Fig. 38.

A 60-year cycle, which we find in the oriental cycle calendar (Seleshnikov 1977), is connected with the synchronization of the periods of revolution of the planets around the Sun. If we express syderic periods of revolution in the Earth's tropical years for Mars (1.880089), Jupiter (11.86223), and Saturn (29.45772), then within a human life a span of 60 years would make a cycle that is close to a period of optimal synchronization of the revolution of these planets.

Consequently, if we take the revolution periods of the planets and the rotation of the Earth as conditions for synchronization, then, considering the boundaries of the unit on development, we can synchronize several main calendars. In this case, the boundaries of the unit of development characterize the sequence of synchronized revolution periods of the bodies in the Solar system.

The sequence of duration of basic boundaries of the unit of development given in Fig. 38 indicates that some of these durations correspond to durations of basic rhythms well manifested in the rhythms of the environment. Thus, the tropical year determines seasonal changes of the Earth, a 22-year rhythm represents a cycle of the solar activity, and a 60-year cycle amply represented in data on solar activity and the processes on the Earth (McCormak and Seliga 1978).

Thus, it might be expected that the sequence of duration values determined by basic boundaries of the unit on development presented in Figs. 37 and 38 characterizes environmental rhythms. In their book Zhirmunsky and Kuzmin (1982) used the unit of development for dating the boundaries of geological history, and, as a result, it was found that the environmental rhythms are synchronized with each other and the changes in the biosphere correlate with them.

According to commonly accepted principles, stages of species formation (phylogenesis) are repeated in the development of the individual (ontogenesis). In

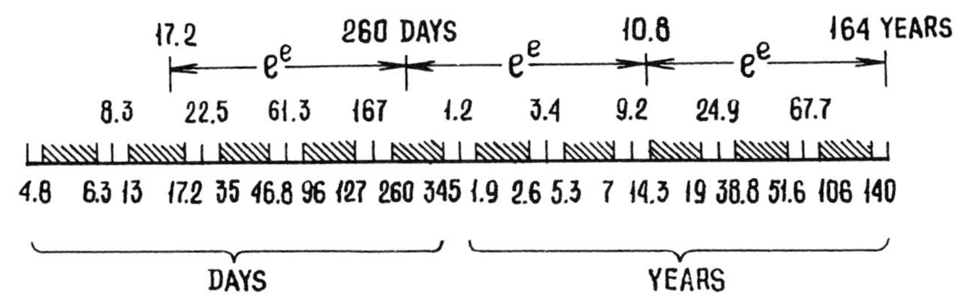

Fig. 39. Sequence of basic boundaries (*end of shaded range*), exponential unstable boundaries (*marks on the lower line*) and reconstruction phases (*shaded ranges*) at $t^K = 19$ years; *above* critical boundaries of stable allometric type

this connection, we can expect that biorhythms, like developmental stages of the biosphere, correspond to the environmental rhythms.

Let us therefore consider data on the development of biological systems. For the analyses we shall use a sequence of basic and exponential unstable boundaries and reconstruction phases of the unit of development (see Table 7) and correlation (72). A corresponding set of boundaries is presented in Fig. 39.

Let us note the synchronization of the critical constants e and e^e, which is needed for further analysis (Kuzmin and Zhirmunsky 1986)

$$e^8 = e^{e^3} .$$

(The correlation of a Turkish cycle type in calendars of $\frac{8}{3}$.) Counting off this relation from the basic boundary of 17.2 days we obtain 17.2 days$\cdot e^8$ = 140 years, and 17.2 days $(e^e)^8 = 164$ years.

4.4 Critical Age Periods in the Individual Development of Animals

Let us analyze data on the dynamics of growth in animals. The analysis is based on a method elaborated by Mendeleev who proposed to present the dependence course as sections of the straight line. In this case, the change of one linear section for the other characterizes the critical point. Studying change in the density of alcohol on its dilution with water, Mendeleev established that at points where the straight line breaks, the solution's structure changes abruptly (Mendeleev 1959). This method was widely used by Schmalhausen (1929, 1984) in the analysis of curves plotted for animal growth in order to divide the process into separate natural periods. The author showed that the rate of exponential growth and the allometric coefficient (power index in allometry, a power dependence) mark distinctly the boundaries of some periods of development between which the process runs smoothly, in an evolutionary way, without essential qualitative reorganizations. The break in the curve characterizes switching-on mechanisms of basic qualitative reconstructions of the system. Schmalhausen found that such

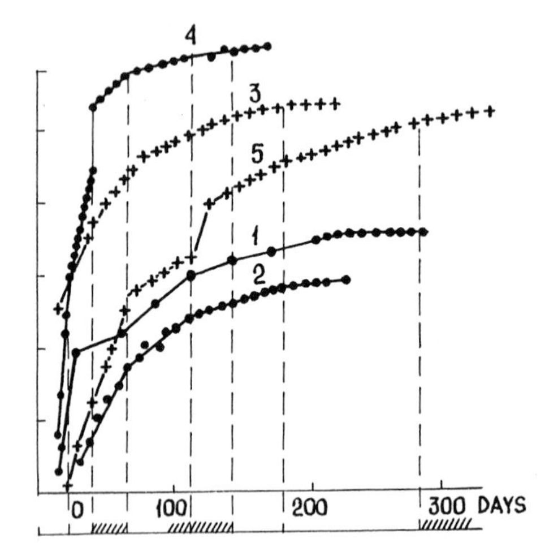

Fig. 40. Dynamics of human embryo mass growth (*1*), postembryonal mass growth for Leghorn hens (*2*) and white mice (*3*), embryonal and postembryonal mass growth for guinea pigs (*4*) and pigs (*5*) (Data from Schmalhausen 1984, V. 1, pp. 158–169) in comparison with boundaries of external rhythms; *shaded ranges* denote reconstruction phases; *marks* show exponential boundaries of unstable type; *above* age in days; *below* boundaries of external rhythms as in Fig. 38

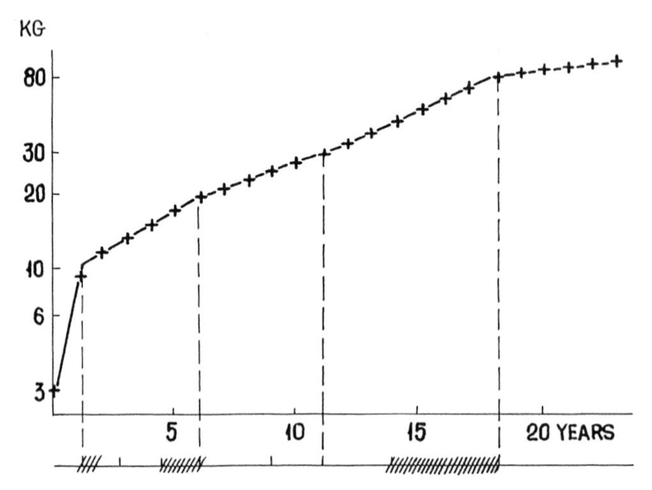

Fig. 41. Dynamics of postnatal mass growth in man and boundaries of external rhythms. Logarithmic scale is used for *y-axis* (Data from Schmalhausen 1984), *shaded ranges* denote reconstruction phases

breaks mark the animal's birth, completion of lactation in mammals, sexual maturation, and ageing.

Let us begin with periodization of the ontogenesis in man. According to data on mass growth in embryonal and postembryonal periods in man (Schmalhausen 1984, vol. 1, p. 75) a graph was plotted of mass logarithm as a function of age (Figs. 40 and 41). In the embryonal period breaks are registered on the 24th, 110th, and 200th day of development with completion of the embryonal period on day 275 (Schmalhausen 1984). The termination of embryo development is an unconditioned and well-studied critical moment in man's ontogenesis. It is believed that under normal conditions the birth of a human child occurs on day 266 after fertilization of the egg cell (Man 1977). Such period duration coincides well with the boundary reconstruction phase in the system of boundaries of external

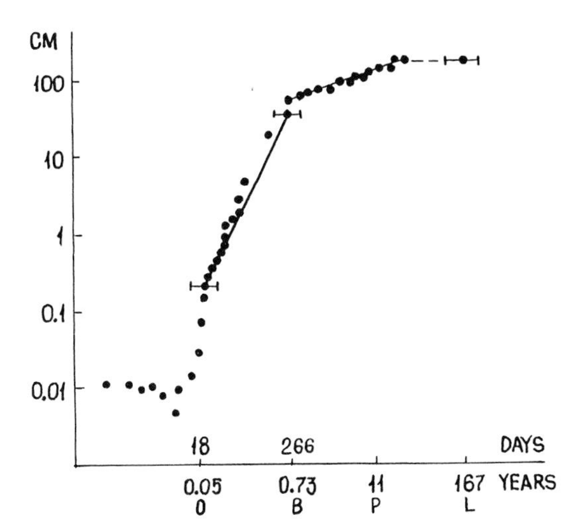

Fig. 42. Dynamics of growth in length for embryo and postnatal stages in man. Logarithmic scales; *0* transition for cellular to organismic period of embryogenesis; *B* birth; *P* puberty; *L* calculated limit of longevity; *Abscissa* age upon fertilization; *Ordinate* length

synchronization rhythms presented in Fig. 37. The previous critical moment determined by a critical correlation with respect to the moment of birth is also well marked. The growth dynamics of the embryo's length and of the postnatal stages in man shows (Fig. 42) that the period of growth with increasing growth rates completes at the age of $266 : e^e = 18$ days. This age in the human developing embryo corresponds to the end of the gastrulation stage (Stanek 1977). Here we observe the transition from the development with increasing rate of exponential growth to allometric development (see Fig. 38; *Biology Data Book* 1964). Probably, this point should be taken as a transitional moment from the cellular to the organismal period of development, when organismal regularity mechanisms start to switch on. In the moment an increased abortion (mortality) of imperfect embryos occurs, comparable to mortality at birth (Svetlov 1960). This age practically coincides with the basic boundary of external rhythms of 17.2 days (Fig. 39). Data given in Fig. 40 record also the boundaries of external rhythms corresponding to the exponential unstable type (22.5 days) and a boundary of the reconstruction phase (about 100 days). At the same time, the break in mass (weight) growth on day 200 of development has no corresponding analog among the main critical boundaries. Possible reasons of this will be discussed in Sect. 5.1.

The plot of postnatal mass growth in man (Schmalhausen 1984) given in Fig. 37 exhibits breaks at the reconstruction boundaries of external rhythms at about 2, 2.5 and 11 years (the boundary of the third hierarchical level e^e since the moment of birth: $226 \cdot e^e = 11$ years) and 19 years (the main boundary; the age given here includes the embryonal period). Thus, reorganizations in the process of mass growth are found to be synchronized with a number of main boundaries of the unit of development. The following critical ages are also well known. Hence, the onset of a reconstruction phase falling on 38.8 years (or, excluding the embryonal period, on 38 years) corresponds to a critical age when the creative abilities of an individual maximally develop and distinct neuro-mental changes occur (Perna 1925). The boundary of an unstable exponential type at the age of 67 years corresponds to the average longevity in developed countries (Urlanis 1978).

The range of critical boundaries e and e^e synchronized by a relation of a Turkish type cycle $\frac{8}{3}$ determines the range of organismal development. The age of 18 days marks the onset of organismal development, and the following critical age ($18 \cdot e^e = 266$ days) determines the moment of birth and, the same relation (266 days$\cdot e^e = 11$ years) further on, the age of puberty in boys (Avtandilov 1973). Puberty in girls occurs earlier, at the age of about 9 years, which corresponds to the unstable boundary. Then follows the third range (11 years$\cdot e^e)^3 = 167$ years). Because of a low degree of approximation for e through the continued fraction $\frac{8}{3}$ we obtain a range of critical boundaries of exponential (19 days$\cdot e^8 = 140$ years) and allometric (19 days$\cdot (e^e)^3 = 167$ years) types.

It may be suggested that this critical moment determines the potential longevity (Zhirmunsky and Kuzmin 1980), i.e., a critical age that may be attained by some individuals (Frolkis 1969). This value is in agreement with estimates made by most competent gerontologists (Hufeland 1796; Bogomolets 1940; Mechnikov 1964; Comfort 1967).

Thus, boundaries of the unit of development associated with external rhythms correlate with the most significant critical moments in human ontogenesis.

A study of the resistance of sea urchin embryos and larvae to high temperatures (Makarycheva 1983) has shown that the percentage of embryos and larvae of a certain developmental level varies unevenly with time. Let us collate boundaries of the unit of development with experimental data (Fig. 43). As a result, one can see that the reconstruction phase falls on a period between the beginning of the embryo hatching from the envelope (p) and the early gastrula stage (e.g.), whence the organismal development starts. Thus, the calculated boundary of the onset of a reconstruction phase correlates with the moment of the embryo's hatching from the envelope and the transition to larval development. Resistance of embryos to high temperatures rises drastically after this boundary and remains as such further on.

Of interest is an analysis of periods in the assimilation of information and the adoption of habits by a child during its development, as introduced by Piaget (1968). The author distinguishes a senso-motor period lasting from the moment

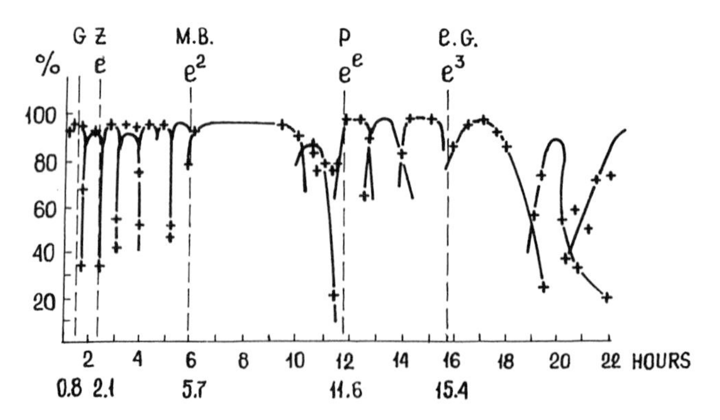

Fig. 43. Course of development of sea urchin embryos and larvae (Data from Makarycheva 1983) and boundaries of the unit of development. *Abscissa* time upon fertilization; *Ordinate* proportion of embryos and larvae of a certain developmental stage

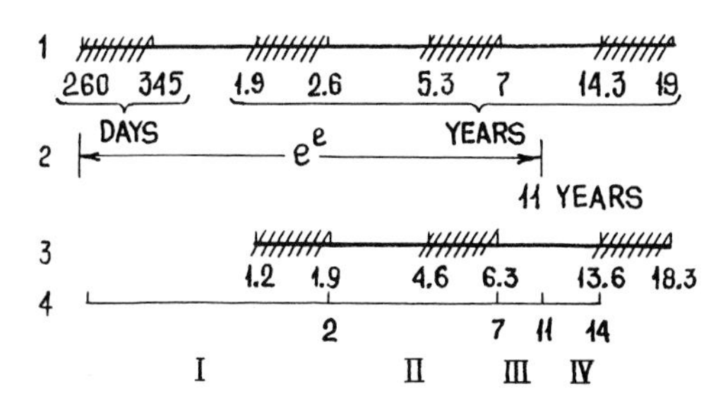

Fig. 44. Periods of acquirement and mastering of information in a child's development (Data from Piaget 1968) and critical boundaries of external rhythms; *1* reconstruction phases of external rhythms, see also Fig. 38; *2* critical boundaries of stable allometric type (e^e); *3* critical ages after birth, years; *4* critical periods, after Piaget; *I–IV* stages of information acquirement and mastering

of birth to 2 years, a period of formation of preoperational mentality from 2 to 7 years, a period of mastering concrete operations from 7 to 11 years, and a period of mastering formal operations from 11 to 14 years. If we shift the unit boundaries by a period of embryo development in order to calculate the age from the time of birth, we shall obtain the sequence presented in Fig. 44. The figure shows that the above mentioned age boundaries, according to Piaget, correlate with the basic boundaries of the unit of development, and an 11-year boundary corresponding to puberty.

Thus, the critical ages of human development are found to correlate with external rhythms of the Solar system, which is reflected in the processes of organism formation and stages of information.

Let us consider data on the embryonal and postembryonal mass growth in chick, guinea pig, pig, on embryonal development of duck, and postembryonal development of sheep, horse, and bull, according to evidence given by Schmalhausen (1984). All these findings are presented in Figs. 40, 45, 46 along with data on human ontogenesis and unit-of-development boundaries characterizing environmental rhythms. With a rare exception (of one boundary for man, chick, and sheep), boundaries characterized by a change in growth trend correlate with critical boundaries of the cell of development for external rhythms.

Thus, the absolute scale of external rhythms determined by basic boundaries of the unit of development characterizes most of the critical ages when qualitative rearrangements occur in developing biological systems. The scale describes rhythms of synchronization in the motions of the Sun, the Earth and the Moon, and determines changes in the characteristics of an environment wherein developmental processes on the Earth take place. Hence, it is natural that essential reorganizations in the development of biological systems, the formation of their developmental stages and the critical phenomena related to morphofunctional transformations take place synchronously with changes in environmental characteristics. In fact, adaptation of biological systems to environmental conditions requires a complete matching of characters, a synchronous functioning of the organism with the environment. Thus, the concept of universal interrelations and interdependence of natural processes finds a quantitative confirmation and manifestation.

Recently, much attention has been paid to the problem of biorhythms, and to the influence of changes in the external factors (climatic and meteorological, in particular). But, as a rule, an illegitimate emphasis is laid on an analysis of single rhythms. For example, 11-year rhythms of solar activity and their effect on the

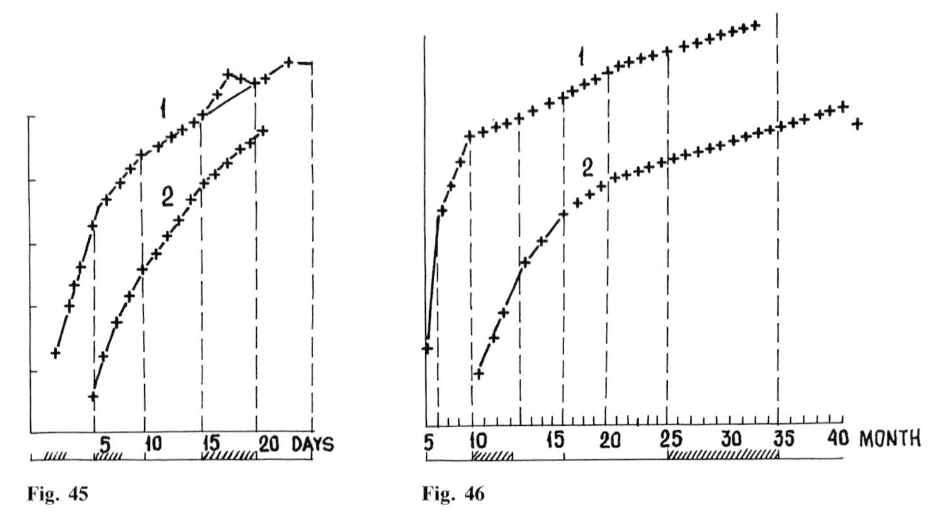

Fig. 45 Fig. 46

Fig. 45. Dynamics of mass growth for embryos of duck (*1*) and hen (*2*) and boundaries of external rhythms. *Abscissa* embryo age, days; *Ordinate* logarithmic scale; *shaded ranges* denote reconstruction phases; *marks* exponential boundaries of unstable type (Data from Schmalhausen 1984, V. 2, p. 62; V. 1, p. 74)

Fig. 46. Dynamics of postnatal mass growth in sheep (*1*) and bull (*2*) and boundaries of external rhythms. *Abscissa* age, lunar months (28 days each); *Ordinate* logarithmic scale, in relative units (Data from Schmalhausen 1984, V. 1, p. 171)

biosphere were rated a great success, owing to the works of Chizhevsky (1976), the founder of heliobiology. There are many investigations of circadic rhythms in the functional activity of organisms and their systems. However, a unified system of rhythmicity, interrelations between various rhythms, their interactions and a reciprocal influence determining the functional activity of systems, are unfortunately beyond the field of research. There are works summarizing empirical data on rhythms of various duration and aimed at constructing rhythm scales. But without theoretical and model-based grounding these results cannot be developed further.

4.5 Rhythms in Human Perception of Acoustic Signals

The question arises whether biological systems of different levels of organization have the same rhythms and, hence, to what extent the results obtained for certain biological objects are applicable to other objects. Nasonov and his co-workers measured the degree of vital staining of preparations of the nerve cells of rabbit spinal cord and frog m.m. sartorii at the action of acoustic signals of different intensity (Nasonov 1962). Figure 47 shows that the maximum degree of staining for both types of cells occurs within the frequency range of 2 to 3.5 kHz irrespective of staining method. The experiments resulted in the conclusion that the effect

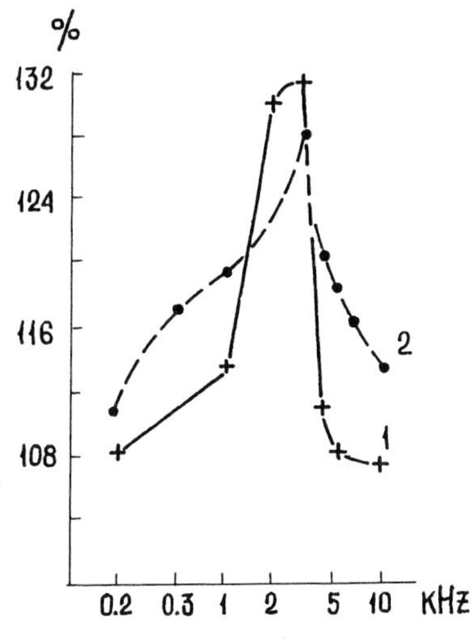

Fig. 47

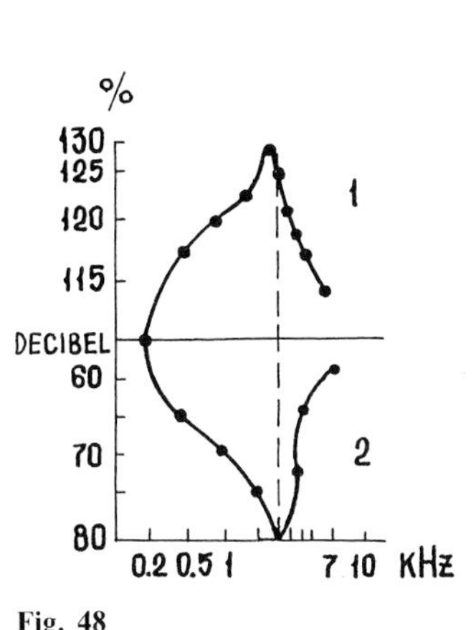

Fig. 48

Fig. 47. Ability of cell staining as a function of acoustic signal frequency; *1* nerve cells of rabbit marrow; *2* frog m.m. sartorii. *Abscissa* frequency, kHz; *Ordinate* fluctuations of staining degree (Data from Nasonov 1962: 33)

Fig. 48. Staining ability of frog m.m. sartorii (*1*) and thresholds of human hearing ability (*2*) as a function of acoustic signal frequency. *Abscissa* frequency; *Ordinate* fluctuation of staining degree (*1*) and threshold of hearing ability (*2*) (Data from Nasonov 1962: 35)

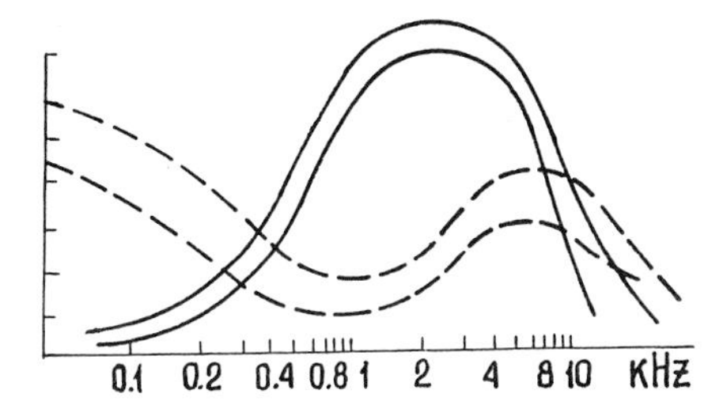

Fig. 49. Degree of speech perception and the level of esthetic response as a function of acoustic signal frequency. *Abscissa* signal frequency; *solid line* denotes the degree of speech perception per frequence band; *dashed line* relative level of esthetic response (Data from Lindsey and Norman 1974)

of a sound with a frequency of about 3 kHz produces essential changes in protoplasmic properties. As the ear tissues contain protoplasm, one may expect an alteration of ear sensitivity within this frequency range. A comparison of the curves for sorption properties of the frog muscles and for hearing thresholds of the human ear (Fig. 48) supported this suggestion (Nasonov 1962).

The dependence of a relative amount of acquired information on the frequency of a sound signal is given in Fig. 49. It shows that the maximum of semantically meaningful information is received in the frequency range of about 2–3 kHz. A shift toward a region of 8 kHz yields a local maximum of esthetic perception. This is a reason for changing the frequency range in radio sets when listening to speech or music (Lindsay and Norman 1974).

Peculiarities in the perception of acoustic signals in the frequency range of 3–4 kHz are physiologically determined too. The external auditory meatus may

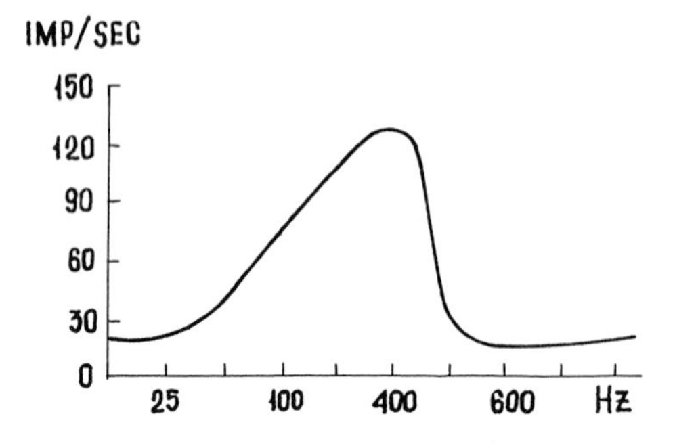

Fig. 50. Tuning curve. See explanations in text. *Abscissa* background frequency; *Ordinate* impulse/sec (Data from Lindsey and Norman 1974)

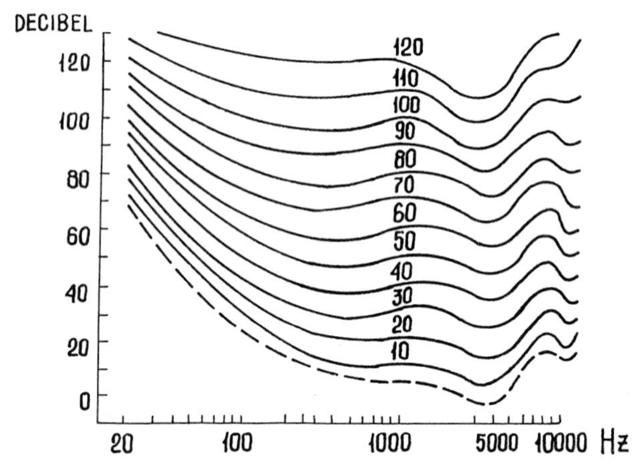

Fig. 51. Isolines and equal audibility at different levels of loudness. *Abscissa* frequency; *Ordinate* sound intensity; *solid lines* show isolines of audibility at a certain loudness level; *dashed lines* denote the threshold of audibility (Data from Lindsey and Norman 1974)

be thought of as a tube closed at one end. The tube resonates when the length of the signal wave exceeds four times the tube length. The length of a human auditory meatus is about 2.3 cm, and, hence, the sound resonance should have a frequency with the wavelength of about 9.2 cm, i.e., about 3.8 kHz. This has been confirmed experimentally by the presence of a smoothed peak of a signal amplitude in the auditory meatus within a range of 2–5 kHz (Gelfand 1981).

It is here that a change occurs in the signal processing regimen, while the auditory nerve is able to respond to signals with the frequency of 3–4 kHz. At frequencies higher than 4 kHz a regular cyclic response of the auditory nerve turns into a chaotic continuous flux of pulses. Therefore, the quality of semantic information transmission at frequencies above 4 kHz falls drastically.

There are critical distinguishing peculiarities in the perception of acoustic signals by man within other frequency ranges too. With a microelectrode introduced into the auditory nerve a background activity up to 150 impulses/sec was registered even in the absence of an external signal. With a pure sound of moderate intensity and high frequency, slowly decreasing, a critical frequency is recorded at which the neuron response increases up to the maximal value (Fig. 50). This dependence is called a tuning curve, because it is the region of a tuning

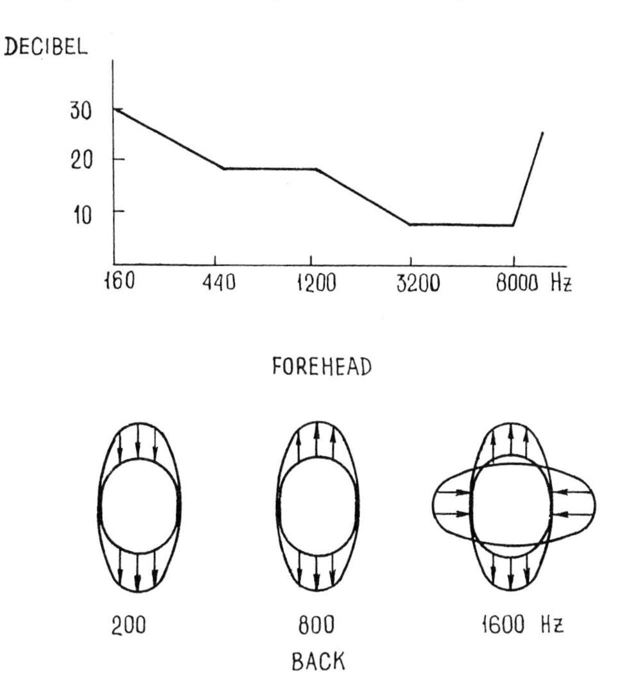

Fig. 52. Relative level of sound effect and vibration of scalp bones at the action of sounds of different frequencies. *Abscissa* sound frequency; *Ordinate* sound intensity; explanations in the text (Data from Gelfand (1984)

fork frequency (440 Hz) for tuning musical instruments. Experimentally established contours of equal loudness (Fig. 51) show that at frequencies lower than 1 kHz the sound intensity is to be increased to maintain a constant hearing level (Lindsay and Norman 1974).

In rhythms of information processing connected with acoustic signals, we can single out a number of frequency ranges at whose boundaries the perception mechanism changes. Gelfand (1981) gives a frequency dependence for relative levels of acoustic effects and of sound transmission through the bone pathway (Fig. 52). It is found that the scalp of man, at the action of sound with a frequency lower than 200 Hz, vibrates as a whole body (Bekesy and Rosenblith 1958). At a frequency of about 800 Hz, the anterior and postanterior parts of the scalp's bones vibrate in counterphase. At a frequency above 1600 Hz, the head bones begin to vibrate as four different portions.

It is to be noted that pleasant voices have their maximum amplitude at frequencies where the perception mechanisms of acoustic signals are known to change. Figure 53 presents the male voice spectrum (Morozov 1977). It is seen that the intensity maxima are evenly distributed and show up the above frequencies (formants). In female voices, three maxima are revealed corresponding to those in male voices, but the region of the main tone, an octave higher, practically coincides with the region of the lower singing frequency. The mean frequency of the female voice is defined more distinctly than it is in male voices, and the high frequency lies in a range between 3 and 3.5 kHz.

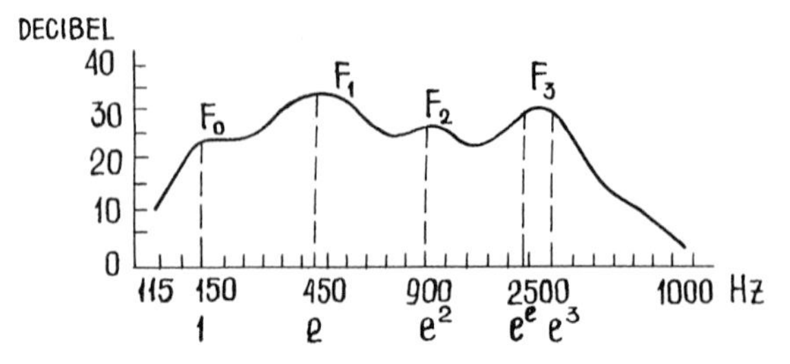

Fig. 53. Integral-statistical spectrum of male voice (tenor), reflecting spectral composition. F_0 basic tone (115–225 Hz); F_1 low singing formants (300–600 Hz); F_2 middle singing formants (700–1300 Hz); F_3 high singing formants (2100–3000 Hz). Low singing formants make the voice soft and "massive", high singing formants make it ringing and clear, silvery, light and melodic. Logarithmic scale (Data from Morozov 1977)

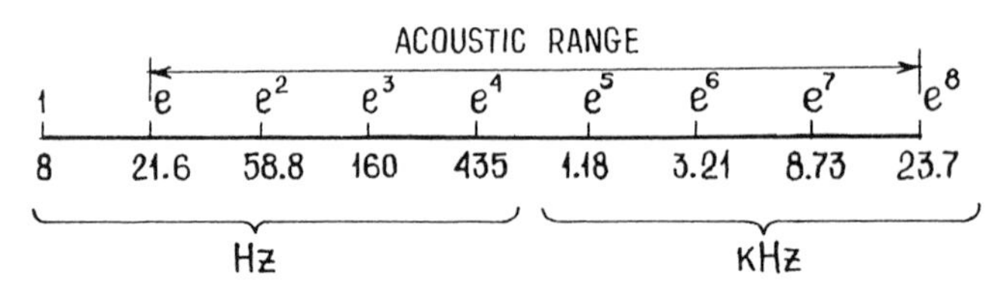

Fig. 54. Basic boundaries of the accoustic range

Perception is reported to be greatly influenced by frequencies which are beyond the range of perception. Human sound perception ranges from 20 Hz to 20 kHz. The periodic sound modulation in amplitude and frequency which, when used correctly, makes the tone color more pleasant, is called vibrato. It is believed that in a human voice vibrato is part of a frequency-controlling mechanism. A vibrato frequency range of 6–8 Hz is optimal for singing. To control the height of your own voice, a feedback is needed, i.e., you need to hear your voice conducted through air or the scalp bones. The time of sound delay in this natural mechanism is about 0.15 sec, which corresponds to the frequency of 6.5 Hz, characteristic of vibrato in a human voice.

The positions of critical values with respect to the unit of development were estimated as follows (Kuzmin 1984). To calculate the position of critical boundaries we will determine the frequency bench-mark. If we count off the synchronization range of critical boundaries of exponential and allometric types $e^{19} = e^{e^7}$ from the period of embryo development taken as a critical boundary $t^k = 260$ days (Kuzmin and Zhirmunsky 1986), we obtain 8 Hz, just the vibrato frequency. Figure 54. gives a sequence of main boundaries of a stable exponential type, which makes after 8 Hz a geometric progression with the modulus e.

Thus, we see that the boundaries where the mechanism of acoustic signals perception changes correspond to basic boundaries of the scale synchronized with the period of human embryo development and external rhythms. As shown in Table 21, some of the boundaries appear to be well known.

The concordance between the position of the main boundaries of the unit of development and the critical frequencies of perception of sound signals would

Table 21. Critical boundaries of man's perception of acoustic signals

N	Hz	Boundary characteristics
0	8	Vibrato, of infra-acoustic range. Adds emotional complexion to signals.
1	21.6	Beginning of acoustic range of 17 – 22 Hz.
2	58.8	
3	160	Background activity of ear at the absence of an external signal.
4	435	440 Hz (la of the first octave) – the basic tuning fork for tuning musical instruments. A sharp increase of ear sensitivity to frequency change.
5	1180	Maximal resonance peak of the middle ear. The local minimum of the perception threshold.
6	3210	The limit of the auditory nerve's ability to follow signals. The absolute minimum of the perception threshold. Resonance frequency of the outer auditory meatus. The maximum of semantic information per frequency band.
7	8730	The local maximum of an amount of esthetic information per frequency band.
8	23700	End of the acoustic range.

seem less striking if one keeps in mind that our ear cochlea is shaped as a conoid spiral with $2^{3/4}$ coils (Gelfand 1981) which means that the human ear cochlea contains e coils.

Consequently, in the work of sensory systems, of the sound analyzer in particular, critical relations of an exponential stable type and boundaries of the unit of development are realized.

5 Critical Levels of the Spatial Characteristics of Systems

This chapter studies a rhythmic pattern of the spatial characteristics of systems. Generally, it appears that the position of temporal boundaries does not always correlate with the spatial arrangement of elements within the system. Specifically, this is true of the position of the Earth in the solar system: this position corresponds to a spatial critical boundary but does not agree with the temporal boundary system.

The existence of preferable sizes of natural bodies irrespective of their physical properties is being widely discussed (Sadovsky 1984). It appears that the rhythms observed in the sequence of preferable sizes agree with critical boundaries of the exponential stable type (e).

The hierarchy of organismic structures and the critical phases of ontogenesis correspond to the critical relation of the allometric stable type (e^e). The mass of the Galaxy nucleus and that of the Sun relate as $e^{(e^e)}$. These results are presented below.

5.1 Distances of Planets from the Sun and the Problem of Basic Variables

The above discussed discretion in the distribution of the system's temporal characteristics leads in some instances to a direct detection of critical values in spatial dimensions. It can be achieved with the aid of the dimensional theory. As it was shown in Sect. 2.1, the correlation between the periods of revolution of the planets of the solar system around the Sun corresponds to the geometric progression with the modulus e, i.e.,

$$T^{k+1} : T^k = e \; .$$

$$T^k = \frac{2\pi (a^k)^{3/2}}{(G M_\odot)^{1/2}} \; ,$$

where a^k is the mean distance of a planet from the Sun, $M_\odot$ is the Sun's mass, $G = 6.670 \cdot 10^{-5}\, cm^3/g \cdot c^2$ (Alven and Arrenius 1979).

Hence, $T^k = A (a^k)^{3/2}$, where

$$A = \frac{2\pi}{(GM_\odot)^{1/2}} \; ,$$

Table 22. Mean distances of planets from the Sun

K	Planet	$\frac{a^k}{a^1} = (e^{2/3})^{k-1}$	Distance from the Sun million km	
			Actual[a]	Calculated
0	Mercury	1	57.85	57.85[b]
1	Venus	$e^{2/3}$	108.11	112.67
	Earth	e	149.6	157.35
2	Mars	$e^{4/3}$	227.7	219.46
3	Asteroids	$e^{6/3} = e^2$	405[c]	427.45
4	Jupiter	$e^{8/3} = e^e$	777.6	832.5
5	Saturn	$e^{10/3}$	1426	1622
6	Uranus	$e^{12/3} = e^4$	2868	3159
	Neptune		4497	
7	Pluto	$e^{14/3}$	5900	6152

[a] From: Childs 1962
[b] Initial value
[c] Mean geometric distance of the asteroid belt from the Sun

which is in conformity with Kepler's law: squared periods of the planets' revolutions relate as their cubed distances from the Sun.

Then a critical relation between the mean distances of these planets from the Sun is obtained from the critical relation for periods of the planets' revolutions ($T^k = eT^{k-1}$):

$$a^{k+1} : a^k = e^{2/3} = 1.94773 \ldots .$$

Positions of the planets around the Sun calculated from this correlation are given in Table 22. The correlation is close to the doubled distances, although redoubling of the planets' distances from Mercury (a planet which is nearest to the Sun) appears to yield obviously overrated results. It follows from Table 22 that successive changes in the distance of planets form the Sun are actually close to a geometric progression with the modulus $e^{2/3}$. Accordingly, critical levels corresponding to the basic rhythm of the unit of development are also revealed in the structure of the solar system. Assuming the distance from the Sun to Mercury, the nearest planet, to be unity, one can obtain a row of values exceeding unity in e, e^2, e^e, e^4 times (Fig. 55). The value as much as e, e^2, e^e and e^4 times greater than the distance between the Sun and Mercury correspond to positions of the Earth, the asteroid belt, Jupiter, and Uranus, respectively. It follows therefrom that there should be an orbit with the critical relation e^3 (between the Jupiter and Saturn), on which a planet might be found.

Thus, a sequence of main boundaries of the unit of development correlates with positions of the planets of the solar system in terms of periods of their revolution around the Sun. As to distances of the planets of the solar system from the Sun, the same unit of development forms new critical boundaries, some of which are synchronized with values obtained from the periods of revolution, and some (e.g., the Earth) form their own critical boundaries. Consequently, in the formation of the solar system, periods of revolution of the planets and their

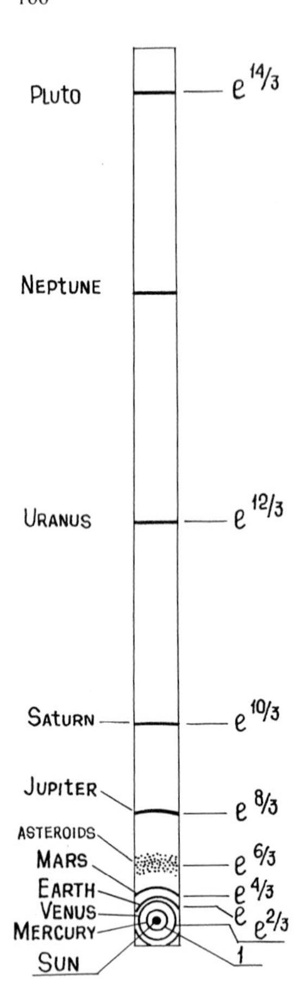

Fig. 55. Relative distances from planets to the Sun. A distance between Mercury and the Sun (57.85 million km) is accepted for a conditional unit

distances from the Sun are the basic variables, as separately they do not define the actual community of planets, but taken all together, they supplement one another in a more complete description of the structure of the solar system.

This kind of interactions between the main series of critical values for various basic variables may result, while analyzing separate properties defined by one basic variable, in seeming irregularities, breaks in regularities of the system characteristics, and arrhythmicity. At such disturbances, however, one should look for relationships whose origin is associated with some basic variables that have not been yet subjected to analysis.

5.2 Discretion of Size Distribution and the Development of Natural Systems[3]

A great number of investigations is at present concerned with the problem of size discretion in natural systems. Sadovsky (1978) paid attention to the phenomenon

[3] This section contains material partly published in a joint paper by V. D. Nalivkin, V. G. Lukyanova and V. I. Kuzmin (Nalivkin et al. 1984)

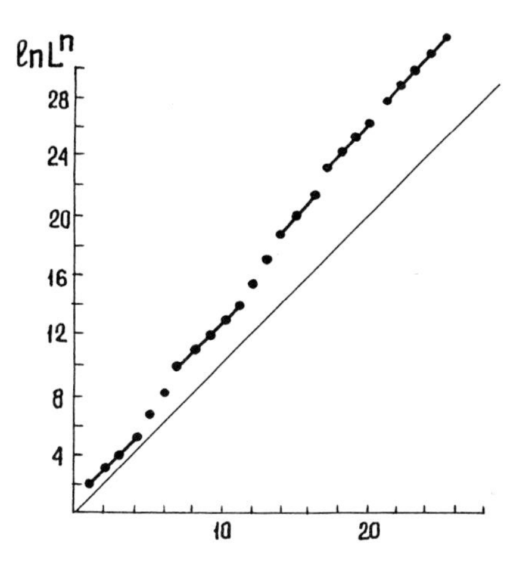

Fig. 56. Maximums for the occurrence of linear sizes of geological bodies (Data from Sadovsky 1984). Graphical presentation of data from Table 23

of a natural clustering in the size distribution of separate pieces of hard material. The author showed that natural systems "prefer" certain sizes irrespective of the properties of material from which structural integrities are formed. It appears that explosions of different intensities crush different rock species into fragments most of which are similar in size (Sadovsky et al. 1982). A study of characteristic sizes from micrometers to hundreds and thousands of kilometers showed that correlations between characteristic sizes are more or less constant, varying in a narrow range from 2 to 5 (Sadovsky 1984).

Let us consider the values of characteristic sizes obtained by Sadovsky (1984) from amazingly voluminous experimental material. Figure 56 presents the logarithms of characteristic sizes of separate bodies as a function of their rank order number. The straight line denotes the portions corresponding to the geometric progression with the modulus e.

We used Sadovsky's approach for a study of other geological objects and made an attempt to reveal regularities in the alternations of irregularities in a sequence of their distribution (Nalivkin et al. 1982). The study was based on measurements of the length and width of tectonic platforms. The smallest platforms are about 1 km wide. In order to expand the mesurements we had to use objects differing in their quality: drifting ice and then blocks of sedimentary rocks divided by their cracks. It should be pointed out that any qualitatively homogenous assemblage of natural objects has its limits. Various types of objects should be necessarily involved if one needs to widen the size range. An analysis of variational rows of sizes in a double logarithmic scale showed that the dependencies are expressed by rectilinear sections divided by breaks. Breaks registered for all objects are listed in Table 23. A graphic representation of these dependencies is hardly informative because of a wide variability range of characteristics in large samples with thousands of measurements. Simultaneously, Table 23 presents a sequence of calculated values as a geometric progression with the modulus e.

Distances between the critical values obtained in logarithmic scale are found to be similar (intervals with no evidence are neglected). Every next value exceeds two or three times the previous one. And the optimal value connecting these data together as terms of geometric progression appears to be e.

Table 23. Critical sizes obtained for geological structures and their comparison with values calculated using critical relationship of exponential type (e-fold relationship)

Calculated values	Break position	Type of structure	Mean characteristic sizes (after: Sadovsky 1984)
2.86 cm	2.9 cm	Rock blocks between limiting cracks	2 cm
	4.5 cm		–
7.76 cm	7.5 cm		7.3 cm
–	11 cm		
21 cm	19 cm		15.6 cm
57.4 cm	50 cm		50 cm
1.56 m	– [a]		1.38 m
4.24 m	5 m	Drifting ice	6 m
11.5 m	11.8 m		
–	23 m		
31.3 m	34 m		28 m
85.1 m			
231 m	300 m		187 m
	400 m		
629 m	600 m		500 m
	900 m		
1.7 km	1.9 km	Platform structures	2 km
4.65 km	4 km		
12.6 km	10 – 11 km		12.5 km
	28 km		
34.4 km	40 km		41 km
93.4 km	90 – 100 km		112 km
254 km	200 km		304 km

[a] Not measured

To reveal the presence of such boundaries in distribution of the mass of natural objects, the distribution of oil and gas reserves from provinces to separate deposits in a storage range from 0.3 to 30000 million tons has been analyzed (Nalivkin et al. 1982). The analysis was performed as in the previous case. Most of the breaks occur at definite reserve levels. These levels are next to equal either for deposits or for areas and districts of oil and gas accumulations. The levels of breaks are divided by similar intervals. The resulting sequence of critical levels of reserves as a whole corresponds to a geometric progression with the modulus e.

Thus, we may suggest a regular alternation of critical boundaries in a series of geological objects and processes expressed as a geometric progression with the modulus e. These critical boundaries mostly coincide with the generally accepted classifications. In geological history, they appear to be proximate to the moments of completion of the main folding processes dividing the geological eras. In the distribution of reserves of oil and gas accumulations varying in scope, the boundaries are close to the limits accepted in the conventional classification of deposits in terms of reserve volumes. Boundaries in the platform size distribution are also close to the generally recognized boundaries of the structure orders. However, the results obtained show that banks and local elevations should evidently be subdivided into a great number of orders, since four or five breaks of the curve occur within them. Division of these structures into two orders suggests itself as having existed for a long time.

Size classification of folded structures is at present lacking. To elaborate such a classification, mass measurements of the length, width and amplitude of foldings would be needed. The drawing of dependence curves similar to those considered above could make it possible to establish natural boundaries of the classes.

Remarkably, spatial and temporal characteristics and masses of geological objects form their critical boundaries in accordance with the same correlation of e-fold intervals between consecutive critical values.

Hence, a necessity arises in research aimed at elaborating a unified coordinated system of spatial-temporal-energy domains, wherein natural systems separated by critical boundaries are developing.

Beyond this theoretic direction, the above established regularities may be used in practice for elaborating natural classifications. In this case, the position of boundaries would be defined more exactly than in expert estimates.

Critical levels may also be used as a basis for analysis of other characteristics and properties of natural objects. So, for instance, it may be expected that at critical boundaries of oil- and gasfields distribution in terms of the reserve volume, economical conditions of mining of these deposits would change drastically.

It is a hard problem to find objects with a high critical relation, e.g., $e^{(e^e)}$, where characteristics at two consecutive points differ 3.8 million times. But recently, a review of a paper by Crawford et al. (1985) has been published in "Priroda" (1986, N3, pp. 99–101), where data are given on the mass of structural formation in the center of our Galaxy.

While studying gas motion in the center of the Galaxy, a team of scientists from the University of California (Crawford et al. 1985) suggested the existence there of a compact object $4 \cdot 10^6\,M_\odot$ in mass (an extremely concentrated stellar accumulation of an unusual composition or a giant black hole). These findings were obtained by three different independent methods, which yielded similar results. Our Galaxy, with the position of the solar system, is schematically presented in Fig. 57.

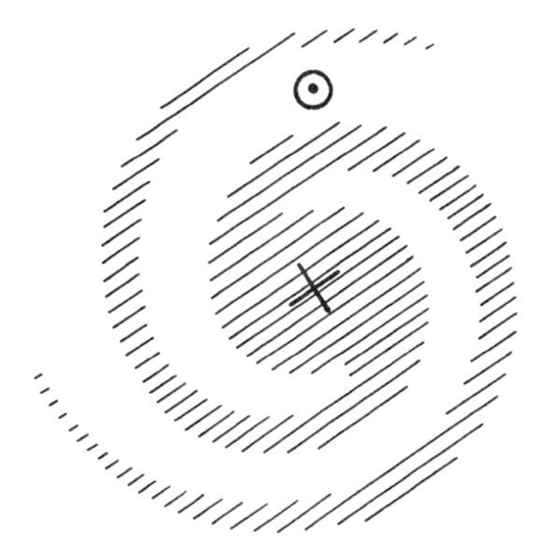

Fig. 57. Schematic presentation of Galaxy; × Galaxy center; ⊙ solar system. *Shaded* "arms" show accumulation of stellar matter (Data from Marochnik and Suchkov 1984)

Thus, the mass of the Sun, $M_{\odot}$ appears to be coordinated with the mass (M) of the central structural formation of the Galaxy through the critical relationship e^{e^e}.

$$\frac{M}{M_{\odot}} = 4 \cdot 10^6 \approx e^{e^e} .$$

5.3 Estimation of Size Ranges of Organismic Structures

As it was noted in Sect. 1.1, classification of hierarchically subordinate biological structures has recently received much attention. Therefore, we shall analyze the relationship between sizes of biological structures of different levels.

The well-known text-book *Cell Biology* (de Robertis et al. 1970) presents a classification of the branches of biology based on size ranges of the objects studied and on the methods of their investigation (Table 24).

The ranges distinguished by de Robertis et al. are shown in Fig. 58, where they are compared with the calculated allometric ranges resulting from successive division of the initial size – 170 cm accepted as the height (body length) of the "conventional man" (*Man: Medical and Biological Data* 1977) – by e^e. $X_2 = X_I/e^e$.

The lower critical boundary of the first range makes 170 cm: 15.15 = 11 cm. Other boundaries are found in a similar way.

In most cases, de Robertis ranges overlap two or more calculated ranges (Fig. 58). This is true for "organs" (3.5 ranges), cells and bacteria (1.5 ranges), cellular components and viruses (2 ranges).

It should be mentioned that de Robertis et al. (1970) note the conditional character of the boundaries between different structures, since these boundaries are established on the basis of the resolving power of the research apparatus used, but not on the size ranges of the biological structures themselves. Besides, the broad variety of data used on structures in animals, plants, and microorganisms

Table 24. Sections of biology (from de Robertis et al. 1970)

Object dimensions m	Section of biology	Structure	Methods of study
$>10^{-4}$	Anatomy	Organs	Visual and simple lenses
$10^{-4}-10^{-5}$	Hystology	Tissues	Light microscopes, X-ray examination
$10^{-5}-2\cdot10^{-7}$	Cytology	Cells and bacteria	Microscope
$2\cdot10^{-7}-10^{-9}$	Submicroscopic (ultrastructural) biology	Cellular components, viruses	Polarizing and electronic microscopes
$<10^{-9}$	Molecular biology	Arrangements of atoms and molecules	X-ray structural analysis

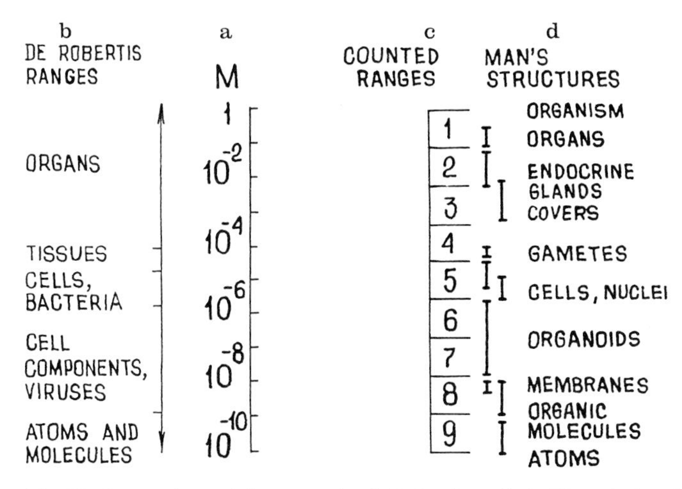

Fig. 58. Comparison of size ranges by de Robertis et al. (1970), calculated allometric ranges and sizes of human structures; *a* scale; *b* range by de Robertis et al. (1970); *c* boundaries and ordinal numbers of calculated ranges; *d* limit boundaries of different structures in the human organism

make for a rather mixed picture that does not allow us to establish clearly distinct boundaries between them.

Hence, taking into account that different organisms may have their own peculiar systems of hierarchy differing in absolute values, we used data on the sizes of structures in man, a rather well-studied species[4], to compare them with the calculated ranges. The data were mostly taken from the reference book *Man: Medical and Biological Data* (1977), text-books on cytology (Troshin et al. 1970) and histology (Eliseev 1972), and some other reference books covering quantitative characteristics of different structures in man and animals (Blinkov and Glezer 1964; Avtandilov 1973; David 1977). These data are given in Table 25 and presented in a summarized form in Fig. 58.

As it follows from Table 25, sizes of organismal structures tend to form definite groups. The first group is the "basic" internal organs (in the table all structures are arranged in accordance with decreasing maximal linear sizes). Among the internal organs, the stomach has the greatest mean size (0.37 m) and the gallbladder has the smallest size (0.1 m) (Fig. 59).

The second group comprises endocrine and some other glands, from thymus (8.4 cm) to epiphysis (7 mm). Lower boundaries of the mean sizes of organs and endocrine glands practically coincide with lower boundaries of the first and second calculated ranges (11 cm and 7.4 mm).

The next sizes cited in the literature are thickness of various walls, coverings, organs and tissues. Here, the walls of the cardiac ventricles (14–7 mm) and stomach walls (13–6 mm) stand sharply out among the others. Then, the derma and subcutis (2.3 mm) follow, and so on. It is difficult to determine the linear sizes

[4] When separate data are available (i.e., referring to men and women), the data on adult men are used. In some cases, when data on man are lacking, material on other mammals is used, and this is then explicitly stated

Table 25. Size ranges for organismic structures in human

No.	Structure	Sizes		Source	
		Limits	Mean value (m)		
1	"Conventional man"		1.7	Man ... 1977	p. 28
	Organs				
2	Stomach	29.5 – 49.5 cm	0.37	Man ... 1977	p. 140
			0.20	Avtandilov 1973	p. 110
3	Lungs	27 – 30 cm	0.28	Man ... 1977	p. 165
4	Liver	20 – 30 cm	0.25	Man ... 1977	p. 153
		16 – 20 cm	0.18	Avtandilov 1973	p. 110
5	Brain	16.5 cm	0.16	Man ... 1977	p. 215
6	Pancreas	14 – 18 cm	0.16	Man ... 1977	p. 157
			0.23	Avtandilov 1973	p. 110
7	Heart	9.7 – 14 cm	0.12	Man ... 1977	p. 125
8	Spleen	10 – 14 cm	0.12	Man ... 1977	p. 114
9	Kidneys	10 – 12 cm	0.11	Man ... 1977	p. 178
10	Gallbladder	8 – 12 cm	0.10	Man ... 1977	p. 155
	Endocrine glands				
11	Thymus gland (3 – 14 years)		0.084	Man ... 1977	p. 114
12	Thyroid gland	5 – 8 cm	0.06	Man ... 1977	p. 210
13	Adrenal glands	3 – 7 cm	0.05	Man ... 1977	p. 206
14	Parotid gland	4 – 6 cm	0.05	Man ... 1977	p. 134
15	Testes	4 – 5 cm	0.045	Man ... 1977	p. 186
16	Prostate	1.7 – 4.7 cm	0.03	Man ... 1977	p. 188
17	Pituitary body	10.5 – 17 mm	0.014	Man ... 1977	p. 210
18	Parathyroid gland	4 – 15 mm	0.007	Man ... 1977	p. 204
19	Epiphysis	5 – 9 mm	0.007	Man ... 1977	p. 208
	Integuments and walls (thickness)				
20	Walls of heart ventricles	7 – 14 mm	$10 \cdot 10^{-3}$	Man ... 1977	p. 126
		2 – 12 mm	$7 \cdot 10^{-3}$	Avtandilov 1973	p. 109
21	Stomach walls	6 – 13 mm	$9 \cdot 10^{-3}$	Man ... 1977	p. 141
22	Skin with subcutaneous fat	0.5 – 4 mm	$2.3 \cdot 10^{-3}$	Eliseev 1972	p. 402
23	Walls of auricles	0.5 – 3.5 mm	$2 \cdot 10^{-3}$	Man ... 1977	p. 126
24	Walls of gallbladder	1 – 2 mm	$1.5 \cdot 10^{-3}$	Eliseev 1972	p. 155
25	Mucous membrane of the stomach	0.5 – 2.5 mm	$1.5 \cdot 10^{-3}$	Man ... 1977	p. 141
26	Skin	–	$1.3 \cdot 10^{-3}$	Man ... 1977	p. 65
	Skin with epiderm	46 – 123 μm	$8.5 \cdot 10^{-5}$	Man ... 1977	p. 59
27	Serous membrane of abdominal wall	0.7 – 1.1 mm	$0.9 \cdot 10^{-3}$	Eliseev 1972	p. 533
	Sex cells				
28	Ovocyte	130 – 150 μm	$14 \cdot 10^{-5}$	Man ... 1977	p. 19
29	Egg cell	89 – 91 μm	$9 \cdot 10^{-5}$	Man ... 1977	p. 192
30	Spermatocyte	58 – 67 μm	$6.3 \cdot 10^{-5}$	Man ... 1977	p. 20
31	Spermatozoon	–	$5.45 \cdot 10^{-5}$	Man ... 1977	p. 187
32	Ovum nuclei	–	$4 \cdot 10^{-5}$	Eliseev 1972	p. 50

No.	Structure	Sizes		Source	
		Limits	Mean value (m)		
	Tissue and blood cells	5 – 40 μm	–	Svenson & Webster 1977	p. 142
				Blinkov and Glezer 1964	p. 382
33	Cortical pyramidal cells	12 – 36 μm	$24 \cdot 10^{-6}$	Blinkov & Glezer 1964	p. 383
34	Ganglion cells of brain	8 – 40 μm	$24 \cdot 10^{-6}$	Blinkov & Glezer 1964	p. 382
35	Mast cells	–	$22 \cdot 10^{-6}$	Eliseev 1972	p. 166
36	Hepatic cells	20 – 25 μm	$22 \cdot 10^{-6}$	Eliseev 1972	p. 513
37	Fibroblasts	–	$20 \cdot 10^{-6}$	Eliseev 1972	p. 161
38	Blood cells	5 – 11 μm	–	Eliseev 1972	p. 131 – 141
	Nuclei of:				
39	Hepatic cells	7 – 16 μm	$11 \cdot 10^{-6}$	Eliseev 1972	p. 513
40	Renal, hepatic and other cells	6 – 14 μm	$10 \cdot 10^{-6}$	David 1977	pp. 28, 30
41	Spermatozoa	–	$4 \cdot 10^{-6}$	Eliseev 1972	p. 50
	Organoids and other ultrastructures				
42	Nucleoli of hepatic cells	1.2 – 2.1 μm	$1.7 \cdot 10^{-6}$	David 1977	p. 40
43	Mitochondria of hepatic cells	0.8 – 2 μm	$1.4 \cdot 10^{-6}$	Eliseev 1972	p. 514
44	Centrioles	0.3 – 2 μm	$1.2 \cdot 10^{-6}$	de Robertis et al. 1970	p. 413
45	Lysosomes	–	$1 \cdot 10^{-6}$	Troshin et al. 1970	
46	Microbodies of hepatic cells	0.4 – 0.8 μm	$6 \cdot 10^{-7}$	Hruban & Rechcigl 1972	p. 27
47	Basal bodies (kinetosomes)	–	$1.5 \cdot 10^{-7}$	de Robertis et al. 1970	p. 413
48	Microvilli (diameter)	0.05 – 0.1 μm	$7.5 \cdot 10^{-8}$	Troshin et al. 1970	p. 86
49	Ribosomes	15 – 35 nm	$2.5 \cdot 10^{-8}$	Troshin et al. 1970	p. 86
50	Microtubes (diameter)	20 – 25 nm	$2.2 \cdot 10^{-8}$	Brown & Rakhishev 1975	
51	Marginated vesicles	15 – 20 nm	$1.7 \cdot 10^{-8}$	Nevorotin 1977	p. 5
52	Desmosomes (intercellular contacts)	10 – 15 nm	$1.2 \cdot 10^{-8}$	Troshin et al. 1970	p. 87
	Membranes in cells (thickness)				
53	Membrane of erythrocytes	8.5 – 9.25 nm	$8.9 \cdot 10^{-9}$	David 1977	p. 136
54	Membrane of oral cavity epithelium cells	7.67 – 9.67 nm	$8.7 \cdot 10^{-9}$	David 1977	p. 137
55	Nuclear membrane	7 – 8 nm	$7.5 - 10^{-9}$	Eliseev 1972	p. 50
56	Membrane of Golgi apparatus	6 – 7 nm	$6.5 \cdot 10^{-9}$	Eliseev 1972	p. 43
57	Membrane of mitochondria	–	$6.0 - 10^{-9}$	Eliseev 1972	p. 41

(Continued)

Table 25 (continued)

No.	Structure	Sizes: Limits	Sizes: Mean value (m)	Source	
	Molecules				
58	Albumin	5 – 15 nm	$10 \cdot 10^{-9}$	David 1977	p. 145
59	Globulin	4 – 12 nm	$8 \cdot 10^{-9}$	David 1977	
60	Actin	–	$5.5 \cdot 10^{-9}$	Finean 1970	p. 128
61	RNA, DNA (step of spiral)	3.0 – 3.4 nm	$3.4 \cdot 10^{-9}$	Finean 1970	pp. 192, 187
62	Myosin (diameter)	2 – 3 nm	$2.5 \cdot 10^{-9}$	Finean 1970	p. 126
63	DNA (diameter of spiral)	–	$2.0 \cdot 10^{-9}$	Finean 1970	p. 186
64	Cholesterol	0.6 – 1.8 nm	$1.8 \cdot 10^{-9}$	Finean 1970	p. 244
65	Polypeptide chain (diameter)	0.98 nm	$9.8 \cdot 10^{-10}$	Finean 1970	p. 119
66	Cellulose	0.476 × 1.828 nm	$4.7 \cdot 10^{-10}$	Finean 1970	p. 298
	Atoms and distances between them				
67	Na atom (diameter)	–	$3.88 \cdot 10^{-10}$	Pimentel & Spartly 1970	p. 56
68	H atom (diameter)	–	$1.42 \cdot 10^{-10}$	Pimentel & Spartly 1970	p. 56
69	Ionic bonds	0.2 – 0.3 nm	$2.5 \cdot 10^{-10}$	de Robertis et al. 1970	p. 58
70	Interatomic distances	0.096 – 0.269 nm	$1.8 \cdot 10^{-10}$	Koptev & Pentin 1977	p. 22
71	Distances between H and O in water molecule	–	$0.98 \cdot 10^{-10}$	Marrel et al. 1968	p. 83

of histological structures proper since some of them spread in "length" throughout the whole body and vary in width very much reaching, in the lower extremes, the size of a cell, e.g., in the monolayer epithelium.

Sex cells, both gametocytes and gametes, fit well the fourth calculated range ($4.9 \cdot 10^{-4} - 3.2 \cdot 10^{-5}$ m). Nuclei of ova are also included here ($4 \cdot 10^{-5}$ m).

An overwhelming majority of tissue cells, blood cells included, have the size of 40 – 5 micrometers (Blinkov and Glezer 1964; de Robertis et al. 1970; Svenson and Webster 1977), which corresponds approximately to the fifth calculated range (32 – 2.1 micrometers). Cell nuclei fall in the same group (Fig. 60).

Organoids of cytoplasm and nuclei (mitochondria, lisosomes, nucleoli), smaller ultrastructures (ribosomes, desmosomes), as well as recently described marginate vesicles, belong to the sixth and seventh ranges ($2.1 \cdot 10^{-6} - 1.4 \cdot 10^{-7} - 9.3 \cdot 10^{-9}$ m).

Depending on functional activity, some ultrastructures change their sizes considerably. This is true for the Golgi apparatus and various vacuoles. Pinocytic vesicles, for example, vary in size between 5 nm to several micrometers (Troshin et al. 1970). Small submicroscopical structures have not yet been studied well. Therefore, it is hardly possible to consider ultrastructures in more detail.

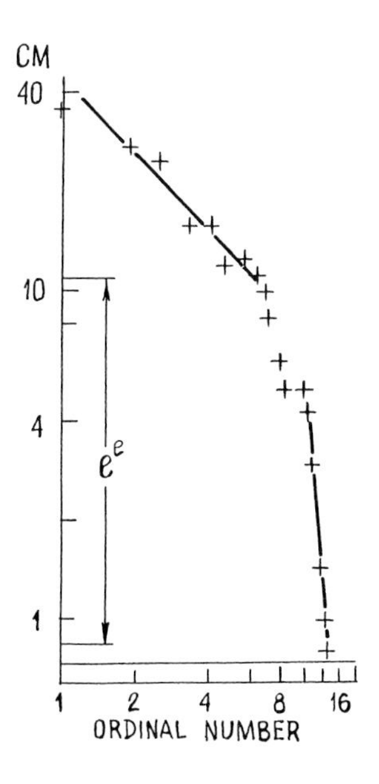

Fig. 59. Variational series of the lengths of the main human organs. *Abscissa* ordinal numbers; *Ordinate* length (lg). Organs are situated in decreasing order of the maximal linear size: stomach, lungs, liver, brain, pancreas, spleen, heart, kidneys, gallbladder, thymus, thyroid, salivary gland, testes, prostate, pituitary body, parathyroid, epiphysis (Data from *Man* 1977)

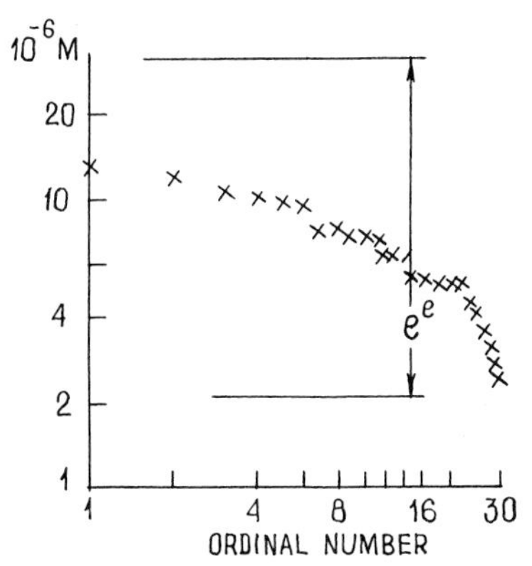

Fig. 60. Variational series of diameters of cell nuclei in different human organs. *Abscissa* ordinal numbers; *Ordinate* diameters of nuclei (lg). Upper and lower boundaries are calculated (The diameters are calculated by means of Khesin's data, 1967)

Next is the group of different cellular membranes whose thickness varies within a very narrow range ($9 \cdot 10^{-9} - 6 \cdot 10^{-9}$ m). Globular proteins, diameters of the molecules of nucleic acids and fibrous proteins and lipids have similar sizes. DNA and protein molecules are much greater. Membranes and organic molecules occupy chiefly the eighth range.

The size of elements of the keratin polypeptide chains (from: Finean 1970) are given in Fig. 61, where one can see a break in the range of $9 \cdot 10^{-10} - 5 \cdot 10^{-10}$ m, which corresponds to the calculated critical boundary.

Atoms are still smaller, the sizes and distances between them correspond to the ninth calculated range ($6.1 \cdot 10^{-10} - 4.0 \cdot 10^{-11}$ m).

A comparison made between the linear sizes of different groups of structures in the human organism and the calculated critical ranges shows that they perfectly

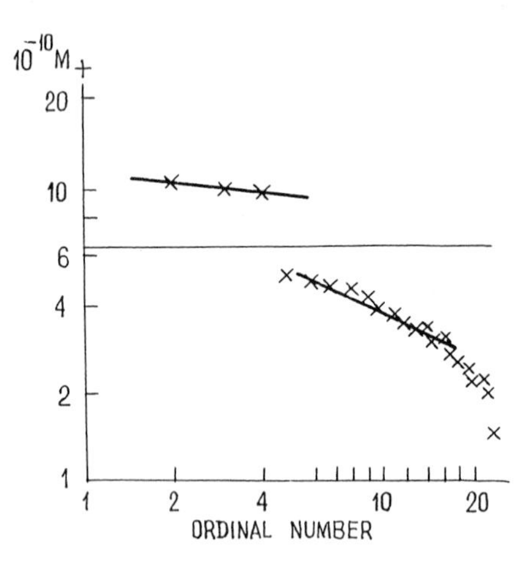

Fig. 61. Variational series of periods of α- and β-keratins at wide angles of dispersion. *Abscissa* ordinal numbers; *Ordinate* periods on x-ray photograph (Data from Finean 1970)

Table 26. Calculated size ranges and corresponding structures in the human organism

No. of the calculated range	Limits (m)		Characteristic structures
	Upper	Lower	
1	1.7[a]	0.11	Main organs
2	0.11	$7.4 \cdot 10^{-3}$	Endocrine glands
3	$7.4 \cdot 10^{-3}$	$4.9 \cdot 10^{-4}$	Integuments and walls
4	$4.9 \cdot 10^{-4}$	$3.2 \cdot 10^{-5}$	Gametes
5	$3.2 \cdot 10^{-5}$	$2.1 \cdot 10^{-6}$	Cells, nuclei
6	$2.1 \cdot 10^{-6}$	$1.4 \cdot 10^{-7}$	Organoids and other ultrastructures
7	$1.4 \cdot 10^{-7}$	$9.3 \cdot 10^{-9}$	
8	$9.3 \cdot 10^{-9}$	$6.1 \cdot 10^{-10}$	Membranes, molecules
9	$6.1 \cdot 10^{-10}$	$4.0 \cdot 10^{-11}$	Atoms

[a] The height of "conventional man" (Man 1977) accepted as initial value for the calculation

agree[5]. This allows us to suppose that the critical ranges serve as a kind of "limits" for size ranges of biological structures. This may be the reason why the structures of the same type are within the bounds of one range (Fig. 58). In the cases where one range is insufficient, a change of dependence occurs which is represented on the logarithmic graph by a break of the line (Fig. 59) and/or a sudden jump in the succession of values (Fig. 61).

Thus, the data considered in this section enable us to differentiate eight ranges, certain organismal structures corresponding to each of them (Table 58). It follows from the analysis performed that the classification of structures based on the methods of investigation (de Robertis et al. 1970) does not agree with their natural hierarchy and, therefore, such classification seems to be inexpedient.

[5] It is necessary to keep in mind that the height of "conventional man" is taken for the initial value, which must be considered as a phenotypic feature. It can indicate only approximately the real mean height, which should be determined starting with the genotypically fixed limits of variability of this feature. But the deviation of conventional man's height from the "real" mean height of man is not likely to exceed 10 cm; and can be considered as satisfactory, particularly as the sizes of organs are given for this very conventional man.

5.4 Critical Levels in Size Characteristics of Animal Ontogenesis

Allometric relationships are widely used in descriptions of interrelations between sizes of biological systems (Huxley 1932; Schmalhausen 1935, 1984; Rosen 1969; Svetlov 1978). This makes it necessary to analyze the essence and character of manifestation of the relation e^e between sizes at critical stages of development. Here, data can be presented either as sizes at critical phases (points) and as allometric functions $y = aX^B$ which change parameters a and B at critical points.

To analyze allometric curves, experimental data are considered in double logarithmic coordinates: for instance, age along the abscissa and size (linear or weight) along the ordinate. This makes a function linear. In other cases, allometric functions with the size (weight) of embryo as the argument can describe, for example, a dependence of metabolism on weight.

An analysis of morphometric data (Avtandilov 1973; *Man* 1977) shows a 15-fold growth of the mass of organs beginning with the start of the postembryonic period up to the cessation of growth. In Table 27, weights (masses) of several basic organs in man are given at the start of the postembryonic period (1.5 months) and when maximal weights are achieved. As follows from the data of the table, the relation of weights at these points is close to critical (e^e). Figure 62 shows the dynamics of growth of lung weight in males. One can see that from the

Table 27. Critical sizes of the main human organs in postnatal ontogenesis (after Man 1977)

Organ	Mass, kg		Correlation (ratio)
	At the beginning of postembryonic development (1.5 months)	Of maximal size	
Lungs	0.069	1	14.5
Stomach	0.10	1.6	16
Spleen	0.015	0.2	13.3
Pancreas	0.0054	0.085	15.7
Heart	0.027 (6 months)	0.34	12.6
Gallbladder (Volume, ml)	3.2	50 – 65	15.6 – 20.3

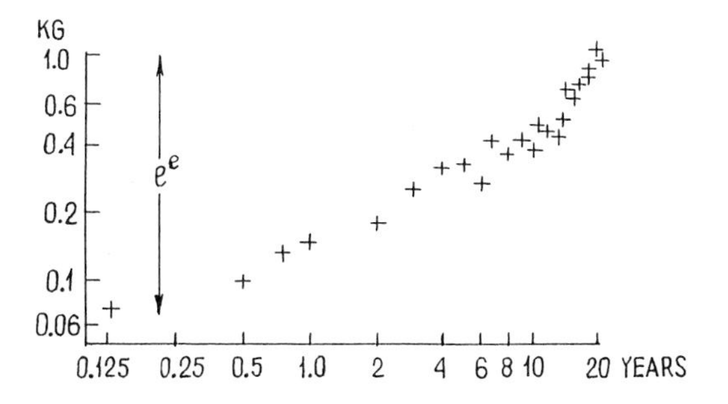

Fig. 62. Age dynamics of lung mass in men in the postembryonic period. *Abscissa* age; *Ordinate* mass. Logarithmic scales (Data from *Man* 1977)

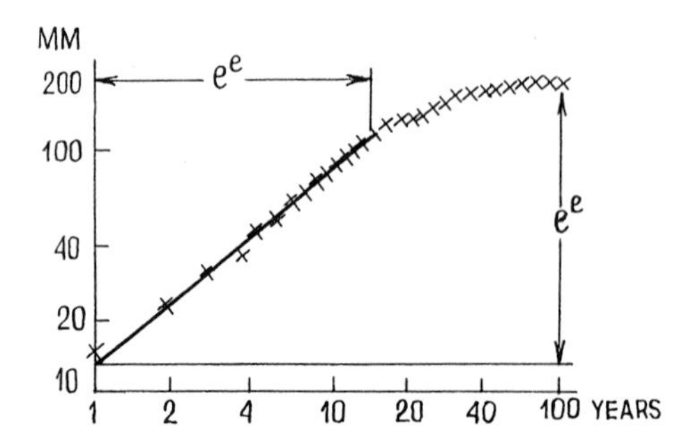

Fig. 63. Age dynamics of the linear size in the mussel *Crenomytilus grayanus* shell. *Abscissa* age; *Ordinate:* shell length. Logarithmic scales

Table 28. Critical shell sizes of the mussel *Crenomytilus grayanus* (after Selin 1980)

Shell number	Age in years	Final size mm	Size at 1 year of age, mm	Relation of final and initial sizes
1	98	189.5	13.1	14.5
2	99	193.1	14.3	13.5
3	95	173.4	13.0	13.3
4	91	201.7	13.3	15.2
5	86	170.2	12.0	14.2
6	89	189.1	13.3	14.2

start of postembryonic development up to the maximal size of lungs at the age of 20 years, e^e-fold changes in weight occur.

Figure 63 presents data of N. I. Selin on the dynamics of growth of the mussel *Crenomytilus grayanus* on stony grounds in Peter the Great Bay, Sea of Japan. Some specimens of this species live as long as about 150 years (Zolotaryov 1974), however an annual increment at an age of $\geqslant 90$ year does not exceed 0.1 mm. Thus, the shell of a 99-year-old mussel whose growth data are presented in Fig. 63 and Table 28 is nearly maximal.

Analogous data for some more shells are given in Table 28. The average geometric relation of final and initial sizes is for them 14.1. Since the mollusc shells have not yet reached their maximal size, this value is slightly underestimated.

The dynamics of a number of nuclei in the rat's body from birth to maximal growth at the age of 50–90 days (de Robertis et al. 1970) showed that these numbers at the start and the end of development relate as $60 \cdot 10^6 : 4 \cdot 10^6 = 15$ (Fig. 64). Thus, in some cases, transition to a next critical level of allometric type can characterize the growth cessation.

The relation of e^e type is frequently found in scientific estimations. For example, Pavlov (1952a) wrote that kidney excretion of metabolic substances can be easily increased about 10 times without any changes in them (pp. 11–12). Zimina (1956) showed that complete adrenalectomy resulted in 36 h in death, but animals with only 0.1 of one adrenal left do not die. Generally, the weight of active tissue is in many organisms 5–15 times greater than the amount necessary for its vital activity.

Figure 65 presents data on the mortality of purebred horses as a function of age (Streller 1964). As one can see from the figure, the range of values within

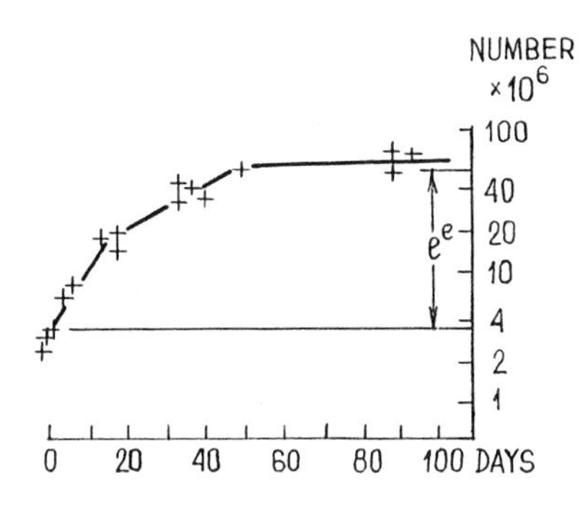

Fig. 64. Age dynamics of the number of nuclei in the rat's body in postembryonic development. *Abscissa* age; *Ordinate* number of nuclei$\times 10^6$ (logarithmic scale) (Data from de Robertis et al. 1970)

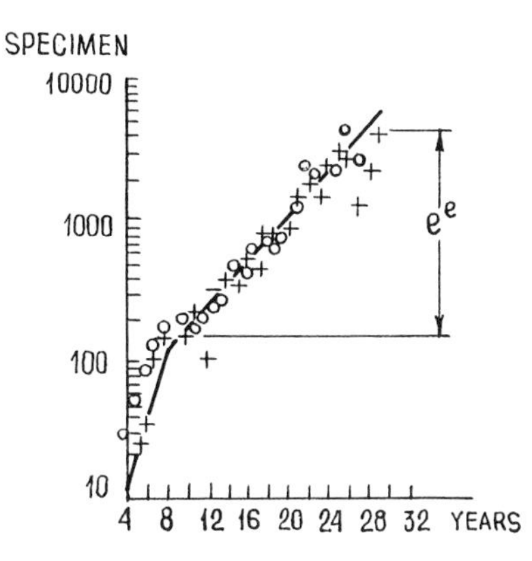

Fig. 65. Change in the mortality coefficient for pure-bred horses, depending on age. *Abscissa* age; *Ordinate* mortality per 10^4 (Data from Streller 1964)

which the tendency of mortality keeps constant has a correlation between critical boundaries e^e.

More detailed quantitative data are available on the embryonic development of some warm-blooded animals. These data characterize the size dynamics. Basic interest in an analysis of size dynamics is explained by the fact that classic research of allometric development (parabolic growth) has been based on these very data (Schmalhausen 1935; Svetlov 1978).

Let us consider qualitative data on the stages of stable allometric development of embryos of warm-blooded animals, that is, age dynamics of sizes for determining critical levels of allometric development. In the given case, the points of the growth curve at which parameters of the allometric curve, i.e., constants A and B of Eq. (3), essentially change, are considered as critical points.

For the curve of growth of embryo sizes in accordance with the function (3), a relation of critical values of ages must be e^e, according to Eq. (56). Knowing one of the critical ages, the end of embryonic development (T), one can calculate the previous critical age: $T^k : e^e = T^{k-1}$.

When the characteristics themselves of the growth curve are unknown, one may use data characterizing intensity of metabolism when the development of an intact organism is being studied. A great number of functions characterizing the interrelations between metabolic intensity and the sizes (weight) of organism of the given species [Eq. (3)] has been experimentally established.

As it was noted before, in these works metabolism was considered as a function of mass, therefore the relation (56) $W_*/W_0 = e^e$ must be valid.

The main conclusion from the above material is that, in connection with the dependence of metabolic intensity on size, the critical relation e^e in the curve of size growth must be observed not only according to the age of the developing system but also according to its sizes.

There are few summarized data characterizing the whole process of embryonic development of warm-blooded animals. The fullest treatment is the section in *Biology Data Book* (1964) where linear sizes are given for embryos of rats, pigs, and man. A series of independent studies were conducted on hens, on their weight characteristics (Schmalhausen 1935; Romanoff 1967; Terskova 1978) and metabolic intensity (Barott 1937). An analysis of the results is presented below.

The allometric curve of embryonic weight growth of chicks (wet weight) from the age of 5 days to the end of embryonic development (from Romanoff 1967) is presented in Fig. 66a. The wet weight at the end of embryonic development is 31.5 g. In conformity with the relation between weights at critical points (56), one may expect the previous critical level at wet weight of $31.5 : e^e = 2.05$ g, which corresponds to the age of 9.5 days. But there is no break at this place of the allometric curve. Let us make a rough estimation of the growth plot using equations

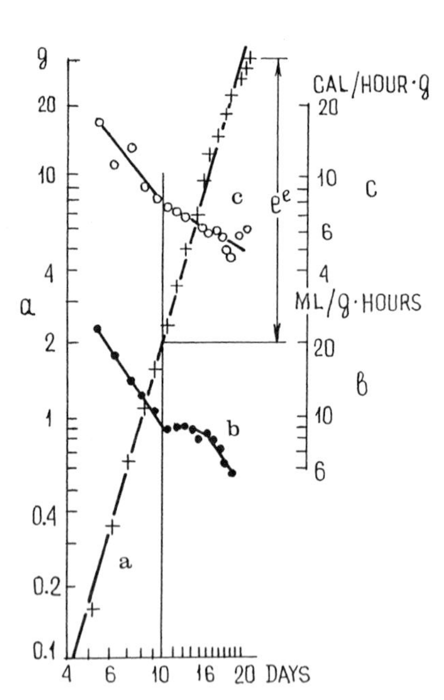

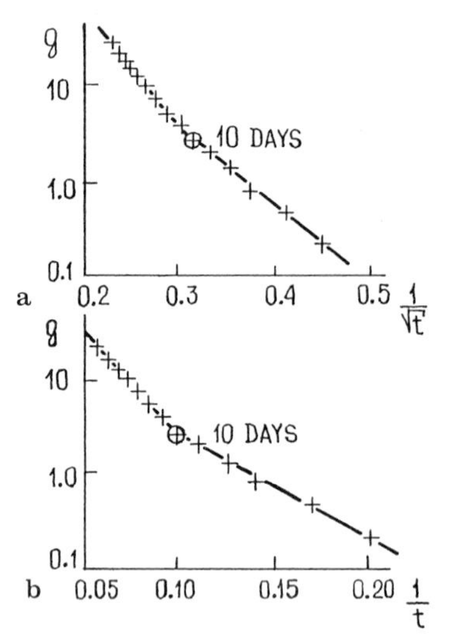

Fig. 67. Change in mass of the chick embryo described by functions (83) – *curve* **a**, and (84) – *curve* **b** (Data from Romanoff 1967)

Fig. 66. The change in mass (**a**), respiration rate (**b**), and thermoproduction (**c**) in the chick embryo, depending on age. *Abscissa* embryo age; *Ordinate a* wet mass; *b* respiration rate, ml/g·hour; *c* thermoproduction, cal/hour·g; e^e the critical range marked off from the hatching moment. The *straight line*, parallel to the ordinate, drawn through the calculated critical point of the curve **a** corresponds to the breaks of curves **b** and **c** (Data from Romanoff 1967; Zotin 1979)

$$\frac{dy}{dx} = B\frac{y}{x^{1.5}} \tag{83}$$

and

$$\frac{dy}{dx} = B\frac{y}{x^{2}} . \tag{84}$$

The corresponding growth curves are given in Fig. 67. It is obvious that the curves, disturbed with regard to the allometric function, show the critical point at about the 9–10th day of the embryonic development of chicks.

This is also confirmed by Terskova (1978), who showed that the curve for the wet weight of chick embryos has a break on the 9th day. This confirms the correctness of classification of the developmental periods of chick embryos (Ragozina 1961): by the 10th day of development a considerable increase of endocrine glands occurs and all temporary embryonic organs (yolk sac, amnion, allantois, serose) reach the maximal size.

Figure 66b gives a curve of change in the respiration intensity of embryos (from Barott 1937) and shows that the value obtained in terms of wet weight for the critical age at hatching corresponds to the change in allometric development of the respiration intensity.

Thus, the curves disturbed in regard to allometric function display a critical point near the 9–10th day of chick embryonic development, corresponding to the critical relation e^e (56).

An analogous break on the 9–10th day occurs on the allometric curve of thermoproduction of chick embryos (Fig. 66c, from Barott 1937).

Let us calculate values of wet weight of the chick embryo for the whole range of sizes, the weight at the moment of completion of embryonic development taken as the initial critical point. Successive critical sizes and corresponding ages are given in Table 29.

Terskova (1978) notes changes in growth rates in terms of weight indices for the chick embryo on the 6th, 10th and 14–15th day of incubation coinciding with sharp morphophysiological changes. Two critical levels from the mentioned three result from the evaluation with the use of the relation between successive critical levels of allometric type. On the 6th day, the period of organ laying is completed, and respiration by means of the vessel area of the yolk sac is supported by allantois respiration (Ragozina 1961). As for data on weight changes in the embryo during the first 5 days, we failed to find any in the literature.

As follows from the theory of dimensions (Migdal 1975), the dimensionless variable at the critical point is of the order of one. In the allometric function (3),

Table 29. Calculated critical levels of the growth of the chick embryo wet mass

Critical level numbers	Wet mass g	Embryo age days
1	28.93[a]	22
2	1.91	10
3	0.126	5.2
4	0.0083	3.5

[a] The initial critical point for calculation – moment of completion of embryonic development (after Terskova 1978)

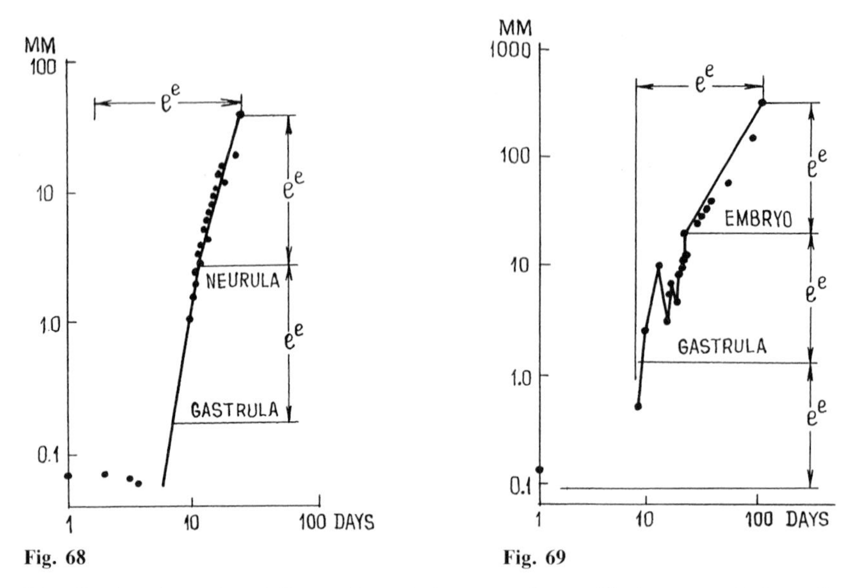

Fig. 68

Fig. 69

Fig. 68. Age dynamics of linear sizes of the rat embryo. *Abscissa* age; *Ordinate* linear size. Logarithmic scales (Data from *Biology Data Book* 1964)

Fig. 69. Age dynamics of linear sizes of the pig embryo. Indications as in Fig. 68 (Data from *Biology Data Book* 1964)

the dimensionless coefficient $b_1 \approx 1$ will correspond to the critical point. This results in the function $W_*/W_0 = x_*/x_0 = e^e$, i.e., among other critical relations of sizes, there must be also critical relations of e^e type. This agrees with the mentioned concept that the relation e^e must be presented in dimensions of developing biosystems, because the dimension is an argument of allometric functions for the intensity of metabolism (Prosser 1973).

We have failed to find so many diverse data for other objects. There are established results characterizing changes of the linear size of embryos with age and embryonal developmental stages. In this connection, an analysis of critical periods of allometric development for these objects will contain only the results characterizing growth curves and allometric dependences between stage duration and stage number.

The allometric curve of linear size growth in the rat embryo (from *Biology Data Book* 1964) is given in Fig. 68. Critical moments of an allometric type calculated by the size value at the end of the embryonic development fall on stages of neurulation and the start of gastrulation. In the period of neurulation, the critical point corresponds to stage 16 (13–20 somites). The previous critical point seems to fall on stage 7–8, i.e., late blastula or early gastrula.

The allometric curve of linear sizes of the pig embryo (from *Biology Data Book* 1964) is given in Fig. 69. Critical periods of an allometric type calculated in terms of size at the end of embryonic development fall on stage 29 (embryo size 18.5 mm = 280 : e^e), stage 9, start of gastrulation (1.2 mm), and the first stages of cleavage (0.081 mm).

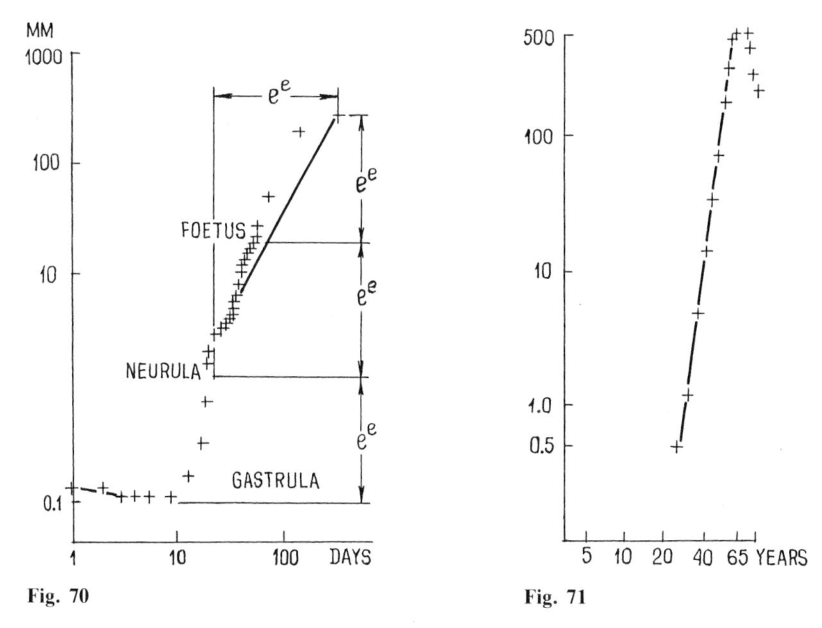

Fig. 70

Fig. 71

Fig. 70. Age dynamics of linear sizes of the human embryo. Indications as in Fig. 68 (Data from *Biology Data Book* 1964)

Fig. 71. Change in the number of men suffering from cancer of the respiratory system in USA depending on age. *Abscissa* age; *Ordinate* number of patients per 10^5 persons. Logarithmic scales (Data from Waterhouse 1974)

The allometric curve of linear size growth of the human embryo (from *Biology Data Book* 1964) is given in Fig. 70. Critical sizes of an allometric type calculated in terms of embryo size at the moment of birth fall on stages of fetus, early neurula, and early gastrula.

An analysis of critical periods at ages of developmental stages shows that critical points fall, as in the previous case, on gastrula (stage 8) and early blastula (stage 1–2).

A similar method can be used to analyze data on the development of systems not connected directly with their spatial characteristics. Data on the number of males in the USA suffering from cancer of the respiratory system as a function of age (Waterhouse 1974) may serve as an example. As one can see from Fig. 71, a common allometric function is observed within the age range from 25 to 65 years. In accordance with Eq. (84), let us roughly evaluate this function, reconstructing therefore the data of Fig. 71 in coordinates ln x, 1/t (Fig. 72). The curve thus obtained has successive breaks at 30, 45, and 65 years, the relations between successive values of the number of sick persons at break points being $32:1.6 = 20$ and $490:32 = 15.3$.

This shows that relationships of an allometric type e^e can be revealed by methods of rough estimation and successively used in the determination of critical ages at which a qualitative change in the character of the developmental process occurs.

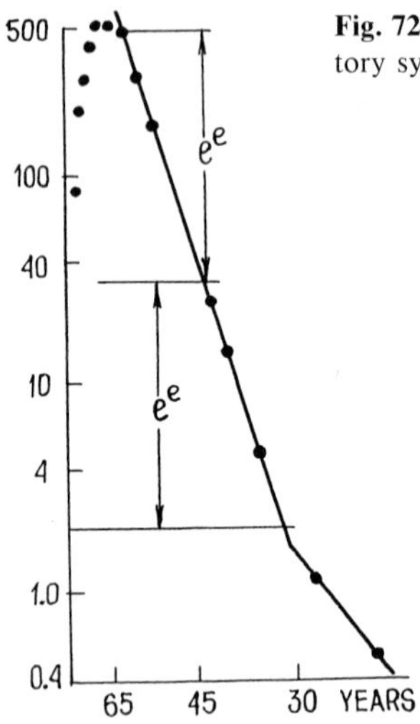

Fig. 72. Change in the number of men suffering from cancer of the respiratory system in USA, depending on age, described by the function (88)

6 Critical Levels in the Structure of Populations and Ecological Systems

An analysis of the structures of natural systems in terms of the manifestation of critical constants is of great interest. Abundances of biological populations and the structure of marine ecosystems determined by density and biomass were chosen as objects of the study. Analysis showed that the basic structure-forming characteristics of these systems are represented by critical relations of an allometric stable type (e^e). It can be expected that more detailed studies will make it possible to reveal also a significant effect of both minor and major critical constants. The results of the study of these systems are given below.

6.1 Studies of Population Hierarchy[6]

Examining how the revealed regularity can be applied to population systems is of considerable interest, since understanding their organization and functioning is important for both theoretical biology and the realization of many practical measures aimed at the protection of animate nature and the management of biological resources. As it is known, population systems are characterized, on the one hand, by fluctuations of the number of specimens the "waves of life" (Chetverikov 1905 and many others) expressed more or less sharply in different species and, on the other hand, by pronounced structure organization and division into hierarchically co-subordinate groups (Shwartz 1969; Konovalov 1971, 1972, 1980; Altukhov 1974; Altukhov et al. 1975; Flint 1977).

Let us analyze the degree of correspondence of data on abundances of population systems actually observed in nature or experiments to the calculated values of critical levels of abundance connected with the reconstruction of a system of allometric type (56).

Consecutive values of critical abundances of populations, in case of their correspondence to a critical relation of an allometric type, will present a geometrical progression with the module equal to e^e. In this case, concrete values of critical population abundances will be determined by the minimal abundance realized. Such abundance, naturally, cannot be less than one. Hence, the theoretical series of consecutive critical values is: $1\,e^e = 15.15$, $(e^e)^2 = 230$, $(e^e)^3 = 3\,500$,

[6] This section contains materials published partly in a paper written together with Prof. A.V. Yablokov (Zhirmunsky et al. 1981).

$(e^e)^4 = 53000$, $(e^e)^5 = 800000$, $(e^e)^6 = 12000000$. At other initial abundances, the relation e^e between consecutive critical levels must persist.

To elucidate the existence of assumed critical levels in the development of populations, it is possible to use the results of direct determinations of abundance of groups within one population complex, taking into account their hierarchical level and population dynamics data.

Direct determinations of population abundance are difficult and hence not numerous, and the accuracy of the quantitative information they contain is often relatively low. Large scales of population systems of a high hierarchical level and the impossibility in a number of cases, to calculate the total of specimens limit the possibility of obtaining precise information of this kind. Consequently, data on the above-mentioned groups can be regarded as mere assessment. However, the results of such investigations, aimed at singling out groups autonomous in a territorial and informational sense, are directly connected with the determination of critical levels in population systems. Therefore, such data are highly interesting for analysis.

Data on population dynamics can be obtained from any sufficiently long-term study of a group. When the development of populations proceeds in conditions where the influence of external factors limiting the abundance increase is reduced (for example, as a result of a prohibition to exploit a population of commercial animals), the reaching of such abundance as corresponds to transition to a qualitatively new hierarchical level in the structural organization must be apparent in the stabilization (or small fluctuations) of abundance over a number of years. It may also be the case that transition to a new quality will be apparent in a change of tendency of the curve of abundance increase (this is characteristic of processes where there is little time for structural reorganization). Data on the dynamics of abundance of permanent populations subject to fluctuations caused by external and internal factors are also of great interest.

If initial theoretical premises about the existence of critical levels of abundance are correct, the preservation of the population system's structure undergoing continuous fluctuations of abundance is possible only when the amplitude of abundance fluctuations is limited to e^e range. To verify this assumption, one can use not only direct data on the abundance of a population at different moments of its existence, but also any comparable data on the relative abundance of groups at different phases of the wave of life, for example data on the commercial catch of animals or population densities over several years.

If critical values of abundance are present, permanent fluctuations over sufficiently long time intervals must have an amplitude not exceeding the boundaries of consecutive critical levels. If the theoretically calculated allometric correlation between critical levels exists, it must occur in experimental data as a ratio of maximal to minimal values in the process of fluctuations.

Data on the abundance of groups of the sand lizard (*Lacerta agilis*) (*The Sand Lizard* 1976) at a site of habitat occupying several hundred hectares were analyzed. The lizards were caught and tagged for a number of years (the total number of specimens ran to several hundreds).

Data on the abundance of groups of red mouse (*Clethrionomys glareolus*) in the Podmoskoviye and Kostromskaya regions were also used (Krylov and

Yablokov 1972; Turutina and Krylov 1979). These materials substantially differ from others in that they came from a comparatively large territory (about 40 hectares) where hundreds of traps were arranged in chess-board fashion over the entire area, so that analysis of the captured animals revealed the biochorological picture of distribution as close to the real one.

Information for the analysis of ranges of fluctuations of abundance in various animals was derived from five fundamental treatises on animal ecology (Allee et al. 1950; Watt 1971; Odum 1971; Dajoz 1972) and all the volumes of the journal *Ecology* for 1975–1976. From these sources we obtained 94 tables and graphs that can be quantitatively decoded and contain absolute and relative abundances of different populations at different moments of population cycles.

To obtain exact data on the abundance of groups of specimens in nature is extremely hard. The methods used are based on the determination of relative numbers (density) or the extrapolation of data obtained from local areas by the "tag- and -recapture" technique or other analogous methods. In a number of cases, errors come to 50% and even in permanent investigations using the most reliable methods they are never less than 5%–10%. This makes one careful about data on abundance of natural groups and helps to bear in mind that a given estimate of abundance may change by at least 10%, and most often by 15%–20%.

To examine the presence of critical levels of the allometric type in data on population abundance, both the approaches mentioned above were used. The first approach is based on an analysis of the results of direct measurements of abundance and consists in the examination of variational curves whose breaks in double logarithmic coordinates characterize change of allometric tendency. In this case, along the ordinate is a series of abundance values in consecutive order by decrease or increase, and along the abscissa are ordinal numbers of corresponding members of the series (see Fig. 73). For small ranges of change of abundance such a technique makes it possible to determine the critical level of abundance, and for larger ranges the ratio of consecutive critical points. In cases where data on population abundance dynamics were available, allometric curves were plotted as is usually done in analyses of growth of organisms or size relationships. Besides, breaks of the allometric curve must characterize critical levels of development (Schmalhausen 1935; Svetlov 1960, 1978). The above-mentioned methods of data processing make it possible to determine both the critical levels of population abundance and their ratios.

The second approach to data processing deals with analysis of the results of inderect determinations of population abundance (on the basis of population densities, catch sizes etc.). Emphasis is laid on the analysis of fluctuation amplitudes in the results of measurements. The maximal value of population abundance is divided by the critical relation e^e, the quotient is plotted on a graph characterizing abundance dynamics, and the result is compared with the recorded values of minimal abundances.

Among natural spatial-genetic (chorological) groups of various ranks, the abundance of populations of the sand lizard has been studied in detail (Zharkova 1973; Baranov et al. 1976, 1977; Yablokov et al. 1976; Turutina and Podmarev 1978).

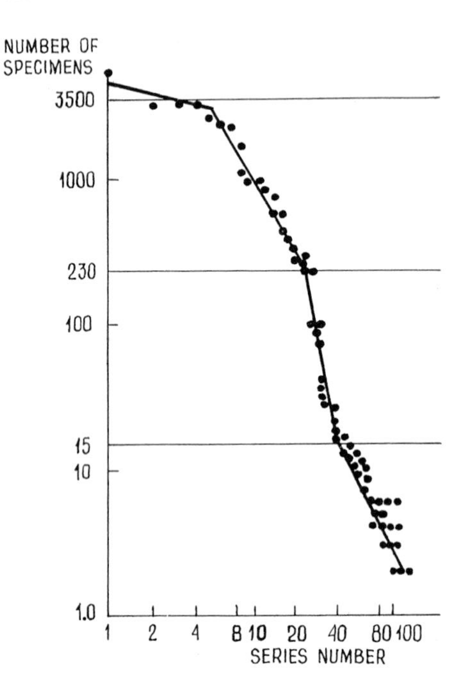

Fig. 73. Variational series of numbers of sand lizard (*Lacerta agilis*) in population groups: *Abscissa* ordinal numbers of successive (in abundance of groups) members of the series; *Ordinate* series of numbers of individuals. Logarithmic scales. *Straight lines* parallel to the abscissa are calculated critical levels (Data from Yablokov et al. 1976)

The abundance of lizard groups, beginning with the smallest groups living in one shelter or in close vicinity to each other and separated from other groups by a certain space, and ending with larger groups distinguished by the presence of spatial isolation and the frequent occurrence of some features – markers of genetic structure (for details see Yablokov et al. 1980), proved to be the following: 2(9), 3(15), 4(12), 5(5), 6(8), 7(3), 8(3), 9(3), 10(4), 11(3), 12(2), 13(6), 14, 15, 16, 17, 18, 19, 22, 23, 26, 27, 28, 30, 33, 34, 38, 70, 85, 98, 99, 220(3), 256, 270, 275, 299, 321, 380, 450, 570, 584, 695, 700, 800, 900, 934, 1050, 1620, 2153, 2250, 2500, 3000, 3240, 3250, 5130 (in brackets is the number of cases in which groups of the given abundance occurred).

Owing to the fact that the abundance of these groups is somewhat underestimated because catching and registration of all the individuals is impossible, Yablokov et al. (1980) drew the conclusion that the number of individuals in groups with minimal abundance (without the offspring of the given year) is in most cases three to five (taking into account that catching is not complete).

Graphic analysis of the given variational curve of abundance of the sand lizard group shows (Fig. 73) that the first break in the curve occurs in the vicinity of 16, the second at 230, and the third at 2800 individuals (or, taking into account 10%–15% of underestimates, 18–19, 250–260, and 3100–3200 individuals).

Consequently, within the sand lizard population four ranges of abundance can be distinguished: $<18-19$, from 18–19 to 250–260, from 250 to 3100–3200, and >3200 individuals. It is clear that the revealed series of critical levels is fairly close to the expected values: 1, 15, 230, 3500 . . .

Comparison of the results from a chorological and phenetic study of the population structure, together with data on radii of individual activity (Baranov et

Table 30. Characteristic features of basic integration levels in specific population of the sand lizard in the western Altai (after Yablokov et al. 1980)

Integration level	Numbers of individuals	Territory, ha	Level of gene exchange, %	Duration of existence, generations
1	Several specimens	0.1	>50	One to two
2	Several tens	Several single ones	~20	Several
3	Several hundreds	Tens	~3–4	Tens
4	Several thousands	Several hundreds	~0.01	Hundreds
5	Tens-hundreds of thousands	Hundreds of thousands	Elusively low	Thousands

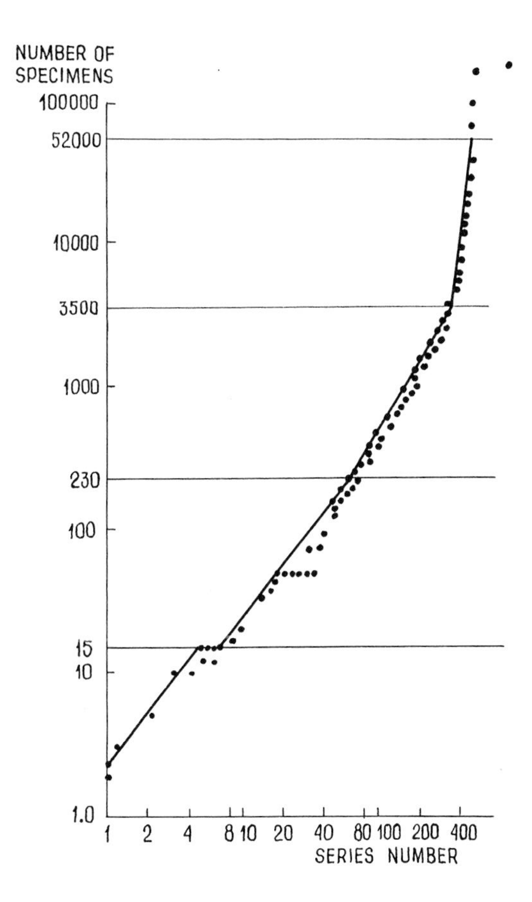

Fig. 74. Variational series of numbers of blueback salmons (*Oncorhynchus nerka*) in population groups on spawning grounds of Alaska rivers. *Axes* as in Fig. 73 (Data from Marriott 1964)

al. 1977), give a general idea of the intrapopulation hierarchy of groups of the sand lizard (Table 30).

From a comparison of the data of Table 30 and Fig. 73 it appears that these data are comparable and complement one another. The exception is the last range of abundance (the 5th level of integration) which corresponds, according to the

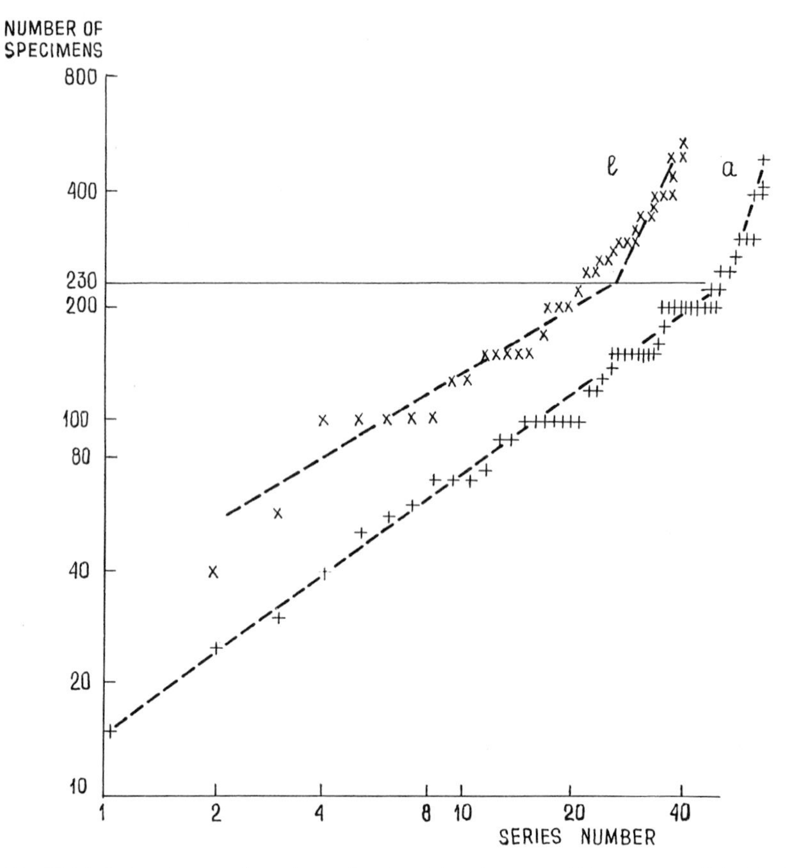

Fig. 75. Variational series of numbers of spring (*a*) and summer (*b*) spawners of the blueback salmon on various spawning grounds of Azabachie Lake (Kamchatka) in 1970–1975. *Axes* as in Fig. 73 (Data from Konovalov 1980)

table data, to two consecutive calculated ranges. If this is true, the sand lizard has six but not five levels of hierarchy.

The second example of a sufficiently accurate determination of abundance of numerous population groups within one species is provided by data on abundance of groups of a Pacific salmon *Oncorhynchus nerka*[7]. In 1946–1962, in the spawning areas of some rivers of Alaska spawners were counted annually with the utmost thoroughness (Marriott 1964). A total of 435 groups were studied. The number of individuals in these varied between 2000 and 160000. From Fig. 74 it appears that there exist critical boundaries of abundance analogous to those described for the sand lizard, where a break of dependences is clearly observed, i.e., the coefficient of allometry changes. These critical boundaries are 10–12, 230, 3000–3500 and ~40000[8]. In Marriott's work (Marriott 1964), no estimates of

[7] We express our gratitude to S.M. Konovalov for drawing our attention to these data.

[8] In Figs. 73–75 curves were plotted using the number series in consecutive order by increase and decrease to show that the result is independent of the method of curve construction. In both cases the results of identification of critical levels coincide.

the precision in calculation are given. From Fig. 74 it follows that there are markedly overstated numbers of individuals corresponding to round figures (50, 100, 200 and 500). However, in a large number of cases estimates were made to the nearest approximations, suggesting that researchers tried to proceed with maximal precision, so that their calculation of salmons in spawning areas with small numbers of individuals (up to 100–150) is precise within ±10% and in case of a greater number of fish it decreases to ±20%–25%, the data in this case being usually understated.

On recalculation of the obtained critical levels (based on interpolation of a smoothing type to compensate for the results of rounding off in the range of abundance of 50 and 100 individuals) the final series of the corresponding ranges of abundance is as follows: 11–13, 275–290, 3600–4000, 48000–50000. It is obvious that the obtained series of values is also close to the theoretical ones: 15, 230, 3500, 58000.

The critical character of the range of abundance of 220–260 individuals is also apparent from an analysis of the variational series of the numbers of blueback salmon in various spawning areas in Azabach'ye Lake (Konovalov 1980, p. 54–55). Similar results with regard to critical range were obtained for two reproductively isolated races of the blueback salmon – spring and summer ones (Fig. 75).

Larger critical boundaries of abundance were found also for blueback salmon. The critical character of the range of abundance of 0.5–1.0 million specimens is evident from a classification of subisolates of the blueback salmon identified by the maximal numbers of spawners (Konovalov 1980) (Table 31). In this classification the limit values differ ten times for three ranges, five times for one, and two times for one range (0.5–1 million). The calculated boundary of 800000 individuals falls into the last range. According to Konovalov (1980), the maximal number of blueback salmon recorded for some isolates of Asia and America is 11 million specimens. This value also corresponds to the calculated critical level of abundance.

The two examples considered above – with populations of the sand lizard and salmon – are evidence for the presence of a certain and largely similar intraspecific hierarchy of critical levels in the abundance of groups.

All the following data deal with not so large ranges of abundance of intraspecific groups. Analysis of these is justified by a search, if not for critical ranges of abundance changes, but for certain points at which the inclination of a segment of the curve changes, i.e., a zone of change of the coefficient of allometry.

Table 31. Maximal abundance of subisolates of blueback salmon (after Konovalov 1980, pp. 37–38)

Abundance	Range of abundance of specimens	Geometric mean of range, thousands of specimens	Calculated critical values, specimens
Very high	1–10 m		11 m
High	0.5–1 m	710	0.8 m
Moderate	100–500 ths	220	–
Low	10–100 ths	32	54 ths
Very low	1–10 ths	3.2	3.5 ths

6.2 Critical Abundances of Populations

To begin with, let us analyze minimal abundances of groups. The real series of numbers of individuals in 50 small groups of adult and semi-adult red-backed mice of the Moskovskaya and Kostromskaya districts (Krylov and Yablokov 1972; Krylov 1975; Turutina and Krylov 1979; Turutina 1979) is the following: 3(4), 4(4), 5(3), 6(5), 7(4), 8, 9, 10(3), 11(2), 12(2), 13(2), 14, 15, 16(2), 18(2), 20, 21, 22(3), 23(3), 25(2), 29. These data show that most groups of this species consist of 3–7 individuals (or 5–14 individuals, allowing for underestimation). The obtained series of values is practically uninterrupted, but it is seen from Fig. 73 that there is a zone in which the inclination of a segment of the curve changes, which corresponds to the range of 11–20 individuals. Taking into consideration that the catch is inadequate (about 25%–50% of error), the critical zone would correspond to 14–30 individuals.

Similar data were obtained for territorial groups of the house mouse (*Mus musculus*) living in the wild on Scockholm, an islet off the English coast (after Berry 1968): 3, 4, 6, 8, 10, 10, 11, 16, 19, 20, 20, 22, 26, 27, 35, 38, 58, 60, 106. The variational curve for this series (Fig. 76) clearly changes its inclination in the region of groups consisting of 16 individuals. Assuming a 20%–25% underestimate in the count of individuals, their number in the zone of change of the allometry coefficient would be about 19–20.

In sand lizard, Turutina (1979) selects a spatial-genetic group of 15–30 individuals. She believes that the average size of the smallest distinguishable population groups of many mouselike rodents is approximately the same, since similar results were obtained for house-mouse, Norway rat, red mouse, and some other animals.

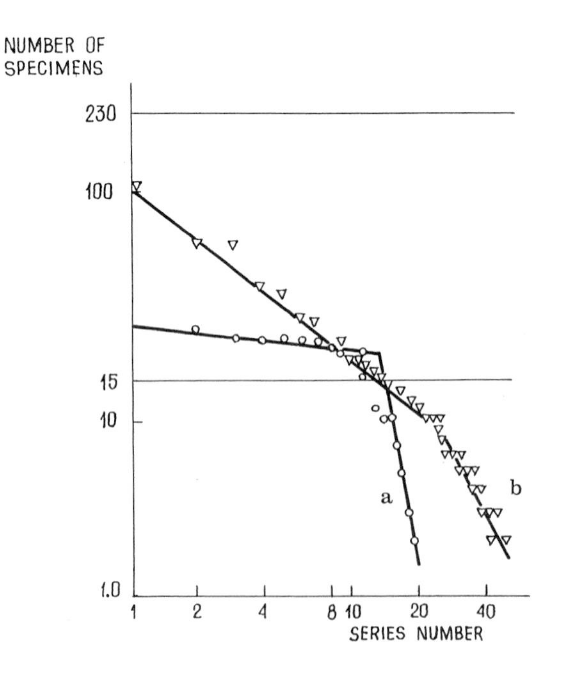

Fig. 76. Variational series of numbers of individuals in small population groups: *a* red mouse (*Clethrionomys glareolus)* (Krylov and Yablokov 1972); *b* house mouse (*Mus musculus*) (Berry 1986) Axes as in Fig. 73

Table 32. Critical levels of abundance of natural groups

Species	Critical values of abundance of groups					Source
	I	II	III	IV	V	
Sand lizard	18–19	250–265	3100–3200			Yablokov et al. 1976, Fig. 73
Blueback salmon	11–13	275–290	3600–4000	48000–50000		Marriott 1964, Fig. 74
Red-backed mouse	14–30					Turutina 1979, Fig. 76a
House mouse	19–20					Berry 1968, Fig. 76b
Black-hat marmot	12–13.5	250	2950			Zharov 1975 (calculation by the authors)
Gray bat			3670	62000		Tuttle 1976 (calculation by the authors)
Bees			2000–3000	50000–70000		von Frisch 1977
Ants				60000–100000	9000000	Zakharov 1978
Average for group	17	264	3200	63000	9000000	
Ratio of average values		15.5	12.1	19.7	14.3	

Zharov (1975) presents data on the abundance of natural groups of the marmot (*Marmota camtchatica*). Of 33 groups of a family level, three groups number 2–5 animals each, five 7–8, and twenty-five 11–16. It has been also shown that 4–7 such families unite into groups of a deme type (44–118 animals, the geometric mean of their numbers being 70). These are followed by colonies of 200 and 300–350 individuals, and three populations numbering 1000–1500, $\geqslant 3000$, and ~ 7000 individuals (Table 32).

Thus, 76% of groups of a family level number 11–16 individuals. These are stable and formed intrapopulation structures (Zharov 1975). The other groups, characterized by smaller numbers only seem to start formation or are represented only by adult individuals. Taking into account the frequency of occurrence, the average number of individuals in groups of the family type is 12 and in stable formed groups 13.5.

Deme type groups consist on the average of 70 animals. The average abundance of colonies is 250 individuals, and the geometric mean of population groups is 2950. Hence we obtain the following series of numbers of individuals in groups of different hierarchical levels: 12–13.5; 70; 250; 2950. The ratio of population group to colony is $2950:250 = 11.8$ and the colony to family ratio is $250:(12-13.5) = 21-18$, with a mean of 15 for these levels. Thus, taking the number of individuals in a formed structure of family level as 13.5, and the correlation between critical levels as 15, the series of calculated values will be: 13.5 (family), $13.5 \times 15 = 200$ (close to the recorded values of abundance of colonies), and $200 \times 15 = 3000$ (the mean size of populations of this species in Zabaikal'e).

Groups with mean numbers of individuals of about 70 (deme type) do not fit in with this series and, obviously, belong to critical levels formed in accordance with mechanisms other than allometry.

Zakhvatkin (1949) presents data on the formation of colonies of phytomonads. The degree of colony integrity is connected with the number of constituent cells; it is characterized by the following categories: (1) less than 16 cells – no signs of integration, all cells are equivalent, after disintegration of the colony each cell gives rise to a new colony; (2) colonies of 16 cells – labile integration, antero-posterior polarity is morphologically expressed, ability to disintegrate persists, each cell gives rise to a new colony; (3) when the colony has 32 cells integration is irreversible.

Ognev (1931) notes that even such an active predator as the pine marten forms during the years of migration of squirrels groups of 10–15 individuals, which then prey on squirrels.

Comparatively precise estimates of the abundance of groups are the results of calculation of individuals in five colonies of the gray bat *Myotis grisescens* (Tuttle 1976). These data include the minimal and maximal numbers: 300–8400, 1600–12000, 16200–19800, 18200–26300, 50000–150000. In Fig. 77, along the abscissa are ordinal numbers of colonies and along the ordinate are numbers of individuals; the upper curve is representative of the maximal values and the lower curve of the minimal ones. It is obvious that the inclination of the curve of minimal values changes in the range of 1600–16200 individuals and that of the curve of maximal abundance in the range of 26300–150000. For the latter series the values from 8400 to 26300 belong to one allometric segment. Therefore, ranges of numbers where the critical points occur are as follows: 1600–8400 and 26300–150000 (Fig. 77, shaded sections). The upper and lower boundaries of these ranges correlate as 26300:1600 = 16.4 and 150000:8400 = 17.9, that is, these values are close to the critical relation e^e. Assuming that the boundaries correspond to the geometric means of these ranges, we obtain the following values: $\sqrt{8400 \cdot 1600} = 3670$ for the first range and $\sqrt{150000 \cdot 26300} = 62450$

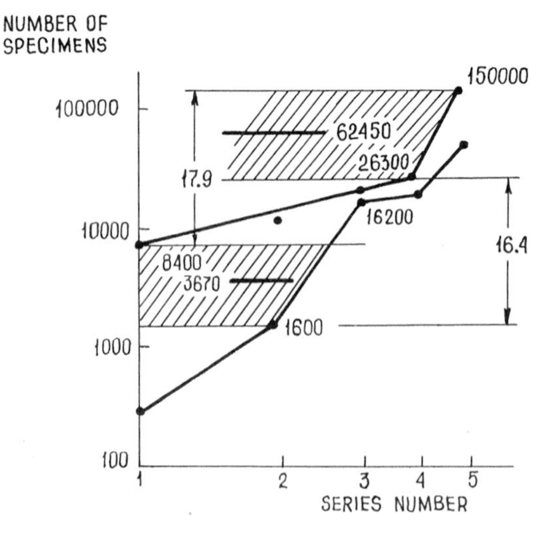

Fig. 77. Changes in numbers of gray bats (*Myotis grisescens*) in five colonies. *Abscissa* ordinals of members of the series (colonies); *Ordinate* number of individuals. Logarithmic scales. Maximal and minimal numbers of individuals are given for each group. Zones of change of allometric tendency are *shaded. Figures near the dots* indicate the numbers of individuals, and *near the lines* parallel to the abscissa calculated boundaries of geometric means of abundance in zones of change of allometric tendencies. For explanations see text (Data from Tuttle 1976)

for the second range, with a ratio of 62450:3670 = 17. Estimates of critical numbers and their ratio are fairly close to the theoretical critical values (3500 and 53000) for the series with an initial member equal to one.

In the honey bee (*Apis mellifera*), a family reaches its maximal peak during the summer months and consists of 50000–60000 adult individuals (Eskov 1979).

Indirect evidence for the existence of certain critical boundaries of abundance of intraspecific groups is provided by data on the size of branches, nests, etc.

Zakharov (1978) reports that a nest of the northern forest ant (*Formica aquilomia*) usually consists of 900000 individuals – a critical value close to the theoretical one (e^e). The numbers are maintained at the same level, and with an increase in numbers the nest forms branches. Thus, large nests of the small forest ant (*Formica polyctena*) divide into separate columns of about 60000–100000 individuals. These intrapopulation groups are informatively isolated and functionally autonomous. The ratio of the numbers of ants in nest to the minimal abundance of column is 900000:(60000–100000) = 15–9.

In four nests of the termite (*Anacanthotermes ahngerianus*) the ranges of abundance of functional groups (Zakharov 1975) are 920–11700 for larvae, 449–4720 for working ants, 22–134 for nymphs, and 8–144 for soldiers. The total numbers in this case are 1399 to 16698. The ratio of total numbers (12) is close to the theoretical value e^e. At the same time, for corresponding functional groups the ratio of maximal to minimal numbers is: 12.7, 10.5, 6.1 and 18, with a mean of 12.

Such critical correlations may appear to be characteristic not only of numbers of individuals on consecutive levels of the linear hierarchy, but also of the abundance of functional groups on each hierarchical level.

On the basis of the material analyzed above, the following ranges of numbers, characterized as critical, can be tentatively singled out (average values for the group of species analyzed): 17, 264, 3400, 6400, and 900000. The correlations between the consecutive members of this series are 15.5, 12.9, 19, and 14, with a mean of 15.4 (Table 32).

On the whole, it can be concluded that the number of individuals in one of the basic small groups within a population of some vertebrates seems to be close to 15–30 individuals (this conclusion does not exclude the existence of smaller and, apparently, somewhat larger groups of individuals of, in principle, the same rank). Less dense groups (of below 15–30 individuals), apparently, belong to a single range of numbers which characterizes the first level of population organization (a family type in mammals).

Groups numbering some tens to 250–300 individuals form the next level of organization. Based on the data on sand lizard, it can be supposed that this level includes several demes isolated from analogous groups by a few generations and with a possible gene exchange of about 20%.

Groups of 250 to 3000–3500 individuals should be referred to the next level of group organization. In the case with the sand lizard, this level includes a group of demes characterized by a comparatively low gene exchange with neighboring groups (about 3%–4% per generation); it can exist as a genetically independent unit over a number of generations, differing noticeably from neighboring groups of the same rank in the specific frequencies of features, markers of a genotypic structure.

As regards groups of 3500–50000 individuals, it can be supposed that they are considerably isolated genetically (about 0.01% gene exchange) and maintain their independence throughout the life span of hundreds of generations.

As shown above for sand lizard, the next level 50000–500000 individuals) includes groups which are completely independent genetically and can exist independently over thousands of generations; they are connected with similar neighboring groups by a scarcely noticeable (per generation) gene exchange. For most vertebrates this level of abundance can be designated as a "real" population level. For most vertebrates characterized by an extremely high abundance of species this level may correspond to clearly expressed intrapopulation units.

Thus, analysis of a wide variety of data on the abundance of individuals and their groups definitely shows that there are stable critical values of numbers which correspond to the geometric progression with the initial member equal to e^e.

6.3 Analysis of Data on Population Dynamics

As mentioned above, one of the ways to verify the proposed hypothesis is analysis of data on population dynamics. In this case, the hypothesis must be substantiated by the fact that fluctuations correspond to the range e^e.

The curve of increase of the biomass of yeast in culture (Dajoz 1972) shows that initially retarded growth changes to an allometric one and then (in those culture conditions) biomass is quickly stabilized (Fig. 78). The ratio of biomasses at the beginning and at the end of the allometric segment corresponds to the critical correlation e^e.

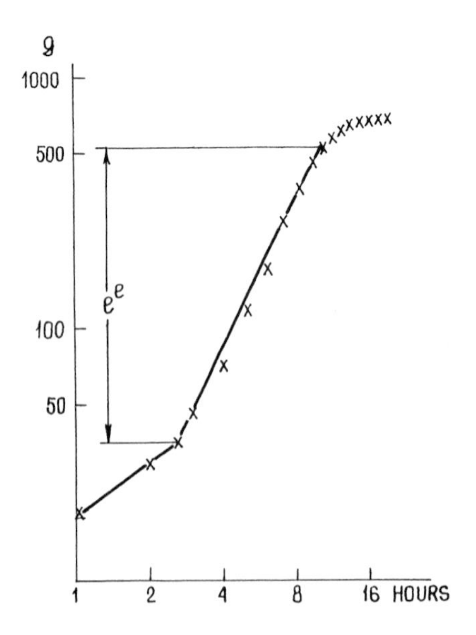

Fig. 78. Increase of biomass of yeast in culture. *Abscissa* time; *Ordinate* biomass of yeast (Data from Pearl 1924)

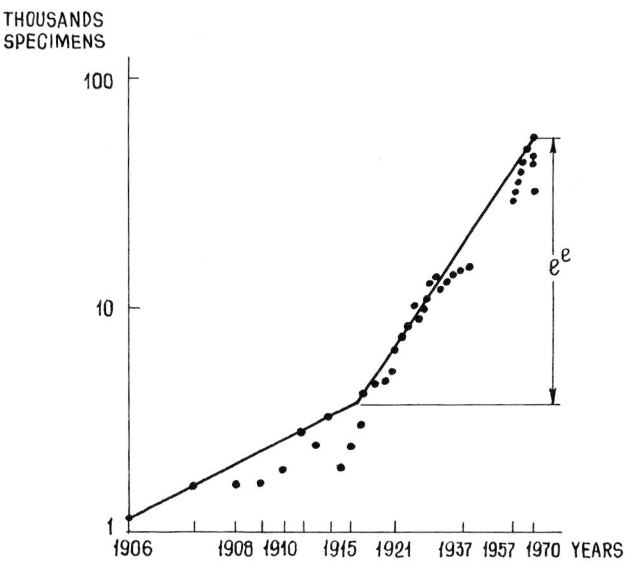

Fig. 79. Changes in numbers of fur seal (*Callorhinus ursinus*) offspring on Tyulenii Island (Sakhalin Island) in 1906–1970. *Abscissa* time; *Ordinate* number of offspring. Logarithmic scale for both axes. *Straight lines* parallel to the abscissa indicate boundaries of critical range e^e (According to Commission on Fur Seal of North Pacific Ocean 1971)

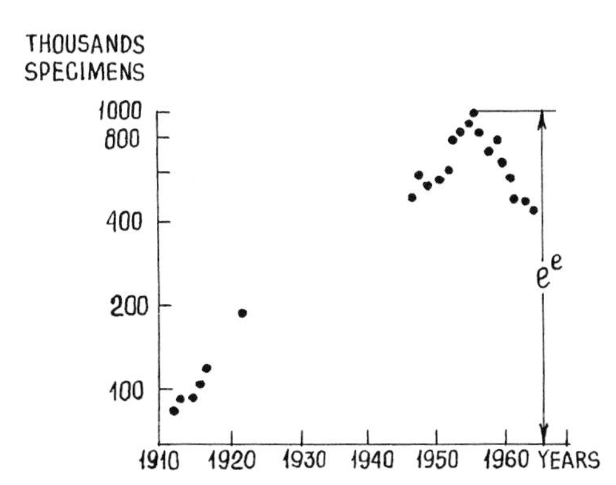

Fig. 80. Change in numbers of fur seal offspring on Pribylof Islands (Alaska) in 1912–1916 and 1947–1966. *Abscissa* time; *Ordinate* numbers of offspring. Logarithmic scale for both axes (According to Commission on Fur Seal of North Pacific Ocean 1971)

An example of intensively growing populations is the growth of fur seal stock after a prohibition of its fishery. Dynamics of the abundance of fur seal offspring on Tyulenii Island during 1906–1970, based on annual records, is given in Fig. 79. It can be seen from the figure that the ratio of numbers for the range characterized by a linear dependence corresponded to the critical relation e^e.

Figure 80 presents data on the quantity of young fur seals born in 1912–1916 and 1947–1966 on Pribylof Islands (Alaska) (Commission of Fur Seal of the North Pacific Ocean 1971). Obviously, the maximal and minimal abundance values (992000, 82000) for newly born fur seals differ 12 times.

One of the classic examples of abundance fluctuations is information about variations in the numbers of lynx (*Lynx canadensis*) and mountain hare (*Lepus americanus*) obtained from 90-year-long observations in the Hudson Bay region. The correlations between the maximal and minimal values in this case are close to 15 (Williamson 1975). Variations in numbers corresponding to e^e were characteristic also of rabbits and coyotes from north-western Utah and southern Idaho (USA) during 1964–1974 (Wagner and Suyde 1978), of thrips (*Thrips imaginis*) living on roses for three years (Davidson and Andrewartha 1948) and of the bug (*Notonecta hoffmanni*) during 1970–1972 (Fox 1975).

A 15–20-fold variation in numbers was noted for the great tit (*Parus major*) in England during 1947–1951 (Lack 1966). Seasonal variations in the numbers of two mosquito species (*Anopheles rangeli* and *Psorophora cingulata*) in Columbia (Bates 1945) were of the same magnitude.

In a number of experiments, abundance variations corresponded to the relation between the maxima and minima of the e^e magnitude: for 20 experimental populations of the fruit fly (*Drosophila subobscura*) 15–20-fold variation in numbers was observed over a 10-week period (Norques 1977). In experiments of Gause (1934) with the predatory infusorian *Paramecium aurelia* and the fungus *Saccharomyces exiguus,* the abundance values differed 15 times. Variations in numbers of the fly *Lucilia cuprina* (Nicholson 1954) over 160 days were also close to 15 times. A study of the interrelations between the typhoid fly *Musca domestica* and the wasp *Nasonia vitropennis* revealed 15-fold variations in the numbers of these species at the first experimental stage characterized by an intensive growth of the populations. After stabilization of the numbers of the host (fly) and parasite (wasp) the stable abundance values proved to differ 15 times (Pimental and Stone 1968).

In a study of the influence of extermination by the predators coyote *Canis latrans* and puma *Felis concolor* on the numbers of the white-tailed deer *Odocoileus virginianus* at the Keibab plateau in Arizona during 1905–1938, Leopold (1943) reported a rise in numbers of the deer to a level differing from the initial one 15 times. Interestingly, death by starvation of the first baby-deer was recorded at the population abundance of about 50000 that corresponds to one of the critical levels of abundance (see above). Commercial catches of the cyprinid fish *Vimba vimba* in reservoirs of Europe during 1950–1967 (Volskis 1973) showed approximately 15-fold variations. Variations in the numbers of squirrel (according to data on the quantity of produced skins) in the Kalininskaya region corresponded to the 15–20-fold range (Tomashevsky 1973).

Thus, the correlation value of 15–20 between the critical levels that corresponds or is close to the correlation between the critical values of the arguments of an allometric dependence, e^e, is fairly frequent in natural and experimental data.

There are a number of examples in which abundance fluctuations are hidden and a special analysis is required to discover them. Let us consider some of them.

The effect of predatory pressure on population dynamics will be analyzed after the example of the sea lamprey *Petromyzon marinus*, the lake salmon *Salvelinus namaycush,* and the lake whitefish *Coregonus lavaretus* in Huron and Michigan Lakes where they had immigrated from the St. Lawrence River in the late 1930's. In Huron Lake the numbers of lake salmon were relatively stabilized at the time of reaching the boundary close to a 15-fold decrease in numbers, as

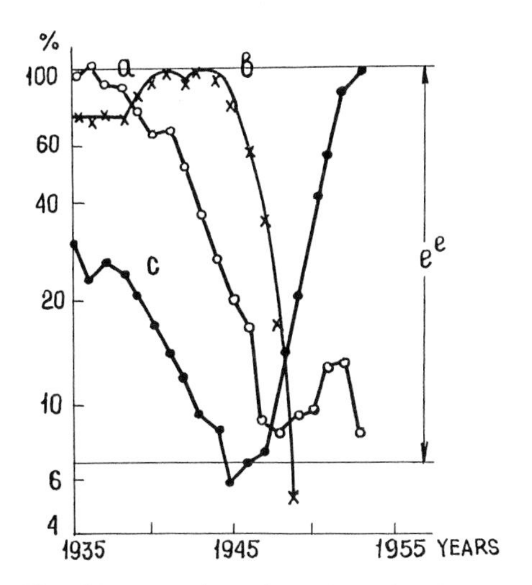

Fig. 81. Dynamics of catches of salmon (*Salvelinus namaycush*) in Huron Lake (*a*) and Michigan Lake (*b*), and of lake whitefish (*Coregonus lavaretus*) in Huron Lake (*c*). *Abscissa* time; *Ordinate* catch; % of maximal catch (100%) over the time interval analyzed. Logarithmic scale for both axes (Data from Watt 1971)

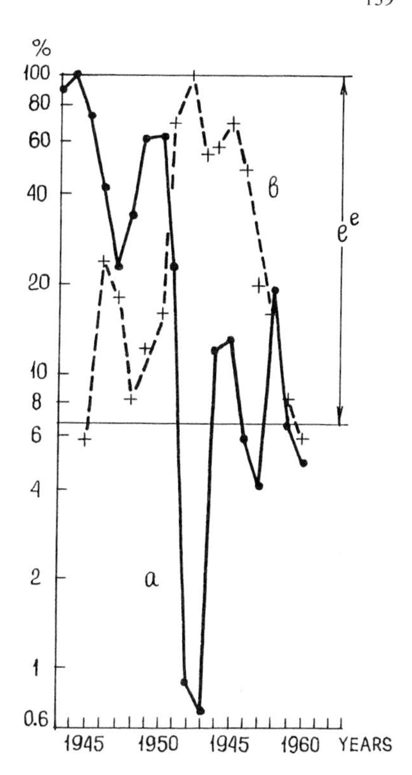

Fig. 82. Dynamics of catches of sardine (*a*) and anchovy (*b*) in Californian waters. *Abscissa* time; *Ordinate* catch; % of maximal catch (100%) over the time interval analyzed. Logarithmic scale for both axes (Data from Watt 1971)

compared with the maximum (Fig. 81). In Michigan Lake, in an analogous situation, the population of the lake salmon disintegrated (Fig. 81 b). At the same time, the lake whitefish in Huron Lake (Fig. 81 c) passed safely this boundary and increased its numbers 15-fold, as compared with the minimum level (Watt 1971).

The dynamics of catch of the sardine *Sardinops sagax coerulea* and the anchovy *Engraulis mordax* off the Californian coast is given in Fig. 82. The catch of sardine fell abruptly between 1951 and 1953 (Fig. 82), while the catch of anchovy (Fig. 82b) rose over the same period (Watt 1971). This is explained by the fact that optimal temperatures for anchovy and sardine are different. The variations in the anchovy catch were of a magnitude of 15–20 times, whereas the sardine catch after an unfavorable period varied over the lower limit of the range of the previous catch.

The influence of fungus infection on the numbers of young herring in the western part of St. Lawrence Bay was elucidated (Tibbo and Graham 1963). The abundance of the herring was estimated according to catch per trawling. Here the catch decreased 15 times, then it rose. This was followed by a new decrease of 15 times, local stabilization, and disintegration of the population.

The dynamics of catch of the northern Baikal whitefish *Coregonus autumnalis* (Kontorin 1980) also corresponded to the range of the critical correlation e^e (Fig. 83).

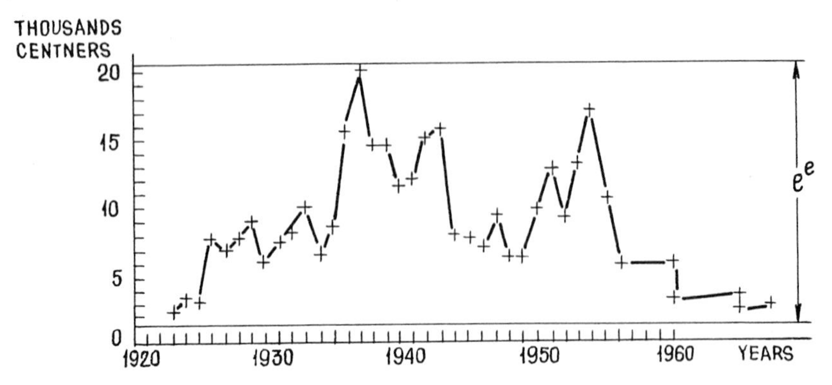

Fig. 83. Dynamics of catches of the northern Baikal lake whitefish *Coregonus autumnalis* in 1923–1967. *Abscissa* time; *Ordinate* catch (Data from Kontorin 1980)

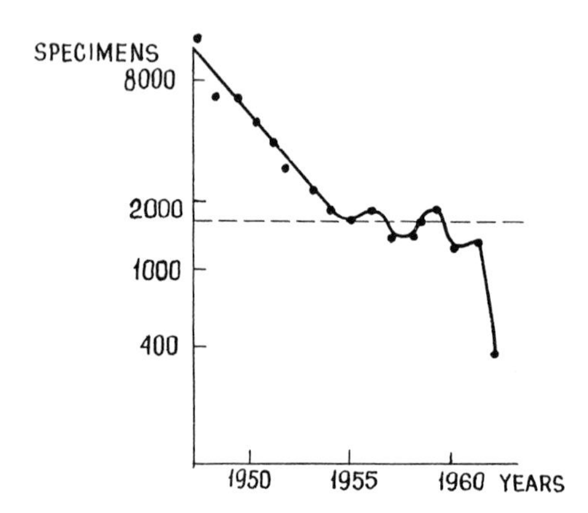

Fig. 84. Dynamics of catch of blue whales in Antarctic waters. *Abscissa* time; *Ordinate* catch. Logarithmic scale (Data from Berzin and Yablokov 1978)

During 1930–1931 in Antarctica 28000 blue whales (*Balaenoptera musculus*) were killed. If we regard this level as the upper limit, then 28000 : 15.15 = 1850 is an estimated value of the lower limit of catch and below this limit there may be irreversible changes of the population structure of the blue whale of the Antarctic Ocean. The fishery practice has fully confirmed this retrospective analysis. As the catch size of about 2000 individuals there was a certain stabilization and if fishery had been stopped at that particular time, in 1960–1961, it might have been possible to renew the population and enhance its abundance to the previous level (Fig. 84). Nevertheless, fishery still more decreased the numbers of the blue whale and created a threat of its complete extermination (Berzin and Yablokov 1978).

Assuming that the initial numbers of the blue whales of the Southern Hemisphere, in accordance with FAO estimates, are equal to 150000–210000, the following levels would correspond to 10000–14000 and 700–1000 individuals. Let us accept that at the time of cessation of fishery for this species in 1965 its abundance was no more than 5000 individuals, then our model allows us to predict a possibility of spontaneous variations in numbers over the range of

700 – 1 000 to 10000 – 14000 individuals. If the population proves to be able to survive when its abundance corresponds to the lower limit of this range then there is, assuming absolute protection, a real hope that the blue whales of the Antarctic Ocean would continue to exist.

Thus, an overhelming majority of the examples analyzed show that there exist, firstly, definite critical levels of population abundance corresponding to the change of the allometric coefficient for variational series, and, secondly, ranges of variations of numbers that correspond to the correlation between the consecutive critical values of arguments of allometric dependence close to $e^e = 15,15\ldots$.

Let us proceed to analyze the abundance of groups and the abundance dynamics of separate populations that disagree with previously supposed propositions. These cases can be classified into three main groups. The first group is the largest in our sample of data and includes cases where the duration of observations is not sufficiently long to reveal the entire possible range of abundance fluctuations. Thus, the abundance of natural populations of the ticks *Asca aphidioides* and *Hypochthonius pallidulus* varied 5 – 6 times over a 5-month period, while the numbers of their host, Collembola, 15.6 times (Ford 1937). The numbers of the grey heron *Ardea cinerea* in two regions of Great Britain changed 2 – 3.3 times over the period of 1934 – 1962 (Lack 1966). The abundance of the common mouse *Microtus arvalis* in the Saratovskaya region (Sigarev and Agafonova 1975) changed only 2.5 times over the period of 1960 – 1971. The numbers of the Californian mouse *Microtus ochrogaster* in one of the populations changed 9 times over 3 years (Birney et al. 1976) and the abundance of the natural population of the house mouse *Mus musculus* on Gall Island changed 10 times over 3 years (Anderson et al. 1964). The abundance of the Arctic fox *Alopex lagopus* population on Yamal Peninsula in 1955 – 1963 changed, judging by the size of the catch, 2.3 times (Smirnov 1967).

Another group of exceptions is contained in data concerning the dynamics of population abundance when numbers can be distorted by anthropogenic effects, for instance, in conditions of control of the numbers of commercial objects by man, in experiments, etc. In such conditions, the full extent of numbers fluctuations may be not manifest. Examples of this are the variations in numbers of two species of flour beetles, *Gnathocerus cornutus* and *Trogoderma versicolor* (2.7 – 5 times) (Park et al. 1941), of the water-flea *Moina macrocopa* (2 times) (Terao and Tanaka 1928), and variations in the catch of the gadid fish *Melanogrammus aeglefinus* in the North Atlantic during 1905 – 1935 (6.5 times) (Allee et al. 1950).

The third group may be the most interesting; it covers the examples of much greater variations in numbers than e^e. The variations in numbers of the larch leaf-roller *Zeiraphera griseana* in central Europe during 1880 – 1960 (Baltensweiler 1964) and of the pine looper moth *Bupalus piniarius* in West German woods during 1880 – 1940 (Varley 1949) were of four orders of magnitude. These variations in numbers seem to reflect real fluctuations, but not the apparent seasonal dynamics of abundance of imago insects which, though extremely significant, is actually nonrepresentative of fluctuations of population abundance because dying adult insects leave the larvae or fertilized eggs for wintering.

For all cases of such drastic large-scale fluctuations, if our theoretical premises are true, the identification of intermediate levels of abundance fluctuations can be expected, while each of them would correspond to critical levels of an allometric type. Verification of this assumption by the available literature material is impossible for lack of data on abundance dynamics and, consequently, it would be extremely interesting to give, in descriptions of such material, information on variations in numbers with time.

Thus, the analysis of deviating data showed that the material of the first group should be excluded from analysis because observations in this case were too short-term and, therefore, could not reveal the limits of abundance fluctuations.

Since the analysis of the relative effects of natural and anthropogenic factors on abundance fluctuations is absent in the works of the second group and it is, therefore, impossible to confirm our assumption on a probable distortion of a natural process under the anthropogenic effect, the second group should be retained for the calculation of the percentage of deviating data. This is also true for the deviating data of the third group, though they seem to characterize fluctuations of another kind.

Above, we analyzed the greater part of data including material on the abundance of groups of two species of vertebrates and nonselected results of the analysis of a comparatively great amount of numerical data and graphs showing absolute values of numbers and fluctuations of numbers for a broad variety of organisms (according to literature sources). Of 94 examples analyzed in the above-mentioned paper (Zhirmunsky et al. 1981), 64 corresponded or were close to the theoretically calculated values and relations (33 of these were analyzed in the above-mentioned paper) and 30 did not (14 if which were analyzed). Eight analyzed examples and 16 remaining discordant examples should not be regarded as contradicting the initial hypothesis, because short-term observations do not cover the entire cycle of population abundance fluctuations lasting sometimes for 20 or more years (see Figs. 80, 81). Thus, 24 out of 94 analyzed examples, in which observations of abundance fluctuations are fairly short-term, should be excluded. Of the remaining 70 cases, 64, i.e., 90%, correspond to the calculated correlation, and less than 10% must be regarded as contradicting it.

The second question concerns structural peculiarities of populations whose abundances belong to one critical range. As follows from the examples analyzed above, maintenance of the abundance level corresponding to the critical one makes it possible to preserve the population structure. In particular, 12–17-fold variations in abundance of biological populations are of the same nature as stationary fluctuations of time intervals. At the same time, factors causing considerable deviations from the given range of amplitudes may result in disintegration of populations or stabilization of numbers at a much lower level. Hence, the correlation between abundance amplitudes corresponding to the critical correlation for allometric system development is fairly stable for populations of principally different biological species. Naturally, this relation is not the only critical one. There exist, probably, smaller and larger gradations of population abundance. However, the relation of numbers e^e is fairly frequent in experimental and natural data, so that it would not but attract attention as one of the essential relations concerning principal changes of structures of intraspecific groups.

The third question arising from the analysis of the data given in this section is connected with the value of the initial member of a population series. When speaking about a theoretical series of critical abundance values represented by a geometric progression with the module equal to e^e as the initial point, we take the abundance value to be equal to one individual. However, it seems very likely that there is a sequence of critical levels with the initial minimal abundance value of two or more individuals. Analyzing the available quantitative data as the allometric ones (in a double logarithmic scale), one can calculate, from the breaks of the curve, the intermediate critical levels of abundance which, in their turn, may be taken as a basis for the calculation of the preceding critical numbers, and so on to the minimal level.

Thus, the critical relation $e^e = 15, 15\ldots$ between the abundance of intraspecific groups of individuals and between limits of fluctuations of population abundance is so frequently encountered that it cannot be regarded as the result of a selection of material or mere occasional coincidence.

Apparently, for the determination of the quantitative limits of existence of various structural unities it would be necessary to carry out special detailed investigations into quantitative patterns of occurrence of groups with different abundance and into the dynamics of population abundance in various species. Of special interest would be a study of the degree of commonness of critical abundances for various species and of a possible mechanism for maintaining numbers on definite structural levels, including also the determination of the functional significance and dominant relations that form structural levels.

These studies would require detailed and more accurate results from the measurements of abundance of natural populations. Unfortunately, such data are at present rather fragmentary.

6.4 Species Structure of Marine Ecosystems

For an assemblage of organisms populating a habitat a number of terms has been proposed, "community" and "ecosystem" being among them. Roughly speaking, the ecosystem consists of a community and a biotope or environment of a given community. The purpose of this section is to analyze the species structure of the community. This question had been raised more than once, but each time the problem was solved mainly by descriptive methods: through studying the community composition that depends on the abiotic and biotic environments and on relations between organisms populating a given ecosystem (Nesis 1977).

Brotskaya and Zenkevich (1939) compared the quantitative characteristics of species composing definite marine bottom communities and detected unevenness in the run of variational curves of a provisory index of density equal to the square root of biomass (g of living matter per 1 square meter) and frequency of occurrence (% of sample plots where a given species was encountered).

$$\text{Index of density} = \sqrt{\text{biomass} \times \text{frequency of occurrence}}.$$

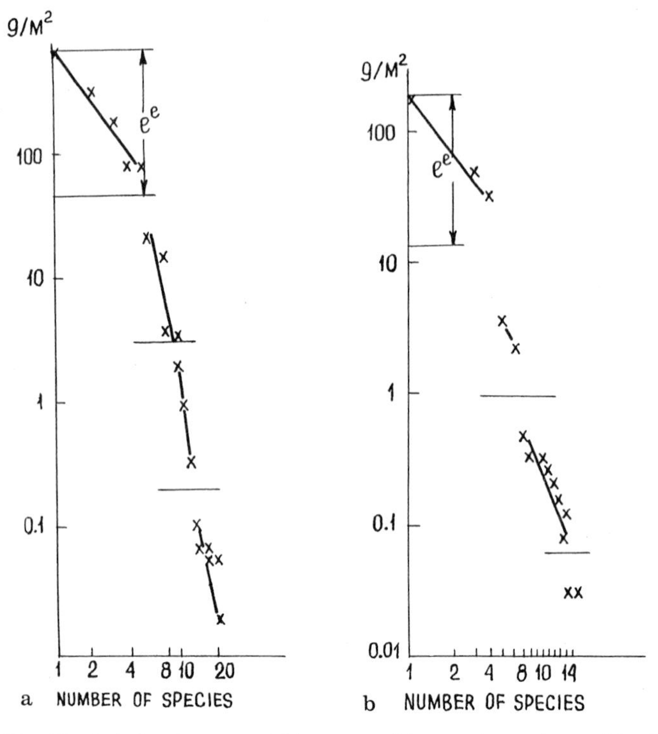

Fig. 85. Variational curves of biomasses of bottom invertebrates **(a)** of the sublittoral (Data from Tarasov 1978), **(b)** intertidal zone (Data from Zenkevich et al. 1948). For explanations see text

Arranging the species included in the definite bottom community of Barents Sea by a decreasing index of density, the authors obtained a step-wise curve. Further, they studied the groups of species divided by steps on this curve and came to the conclusion that in the composition of communities there is a definite structure regularity (p. 82). The authors, however, failed to explain its "causal dependence".

For an analysis, we used data on biomass of species composing bottom and fouling communities (Zhirmunsky and Kuzmin 1982, 1983; Zhirmunsky and Oshurkov 1984). Biomass values of various species were plotted on the graphs. Along the abscissa were numbers of species in a number of decreasing biomasses; along the ordinate were biomasses of species, g/m^2. Logarithmic scale was used for both axes.

The examination of the graphs showed that the biomass values for community species make it possible to draw across them a variational curve distinctly separated into sections (Figs. 85, 86). The points of one section as a rule lie at a common segment whose inclination differs from that of segments joining the preceding and subsequent points. It appears that the points of the next section are distinctly cut off in values from the last points of the preceding section. Thus, an unevenness exists in the distribution of points, a kind of "quantization" in a series of their values.

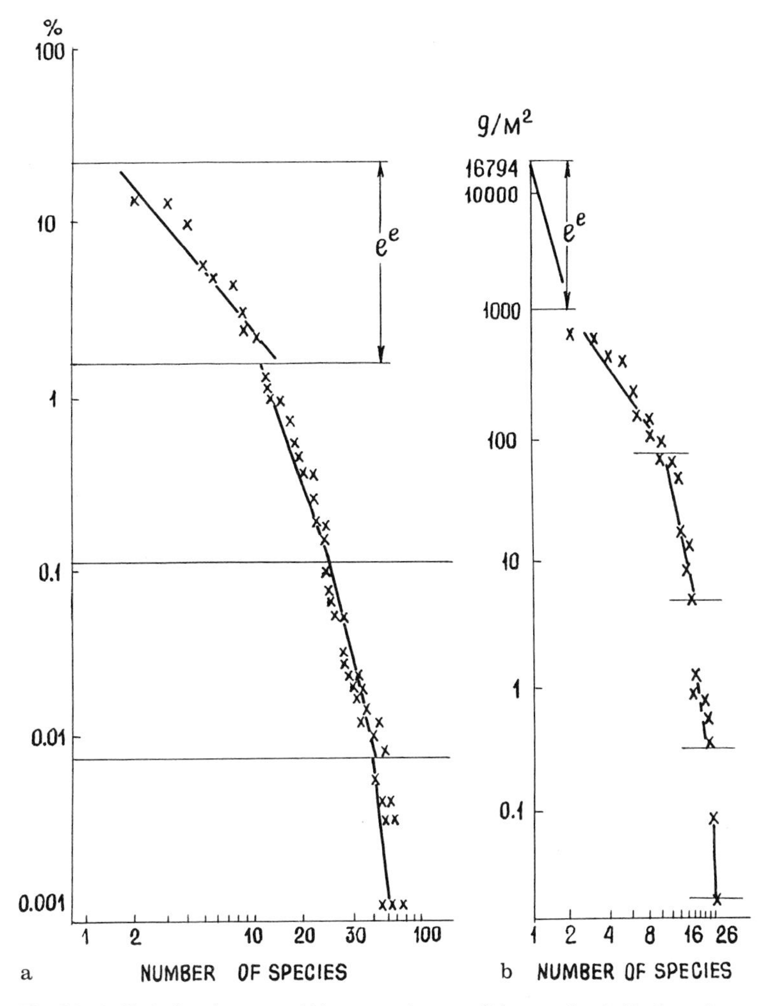

Fig. 86a, b. Variational curves of biomass; **a** bottom fishes on the shelf of north-eastern Sakhalin. *Ordinate* proportion of biomass of a species in total catch (from Batytskaya 1984); **b** species from foulings of experimental plates in White Sea (Data pers. commun. from V. V. Oshurkov). For explanations see text

Figure 85a presents data on biomasses of species in the ecosystem of the bivalve mollusc and sand dollars *Mercenaria stimpsoni* + *Echinorachnius parma* in the open Priboinaya Inlet at depths of 5–10 m. The investigation of this ecosystem was carried out by Tarasov (1978) using scuba diving. The number of registered species of invertebrates was 20 (up to the biomass of 0.01 g/m^2), the total biomass was 1 400 g/m^2, and there were no algae. The variational curve distinctly fell into four groups with breaks between the extreme points and different segmental inclinations.

Let us assume that the "leading" species *Mercenaria stimpsoni* dominating in biomass determines the structure of the community. Further, let us divide its biomass value by e^e and across a value obtained on the ordinate draw a straight line parallel to the abscissa. For a community under analysis such a procedure could be repeated four times. As a result, we obtain four calculated critical levels and four allometric segments between them.

As seen from Fig. 85a, the calculated critical levels fall within intervals between groups of points and thus separate them. Hence, we reveal the species structure of a community composed of several substructures each occupying the range e^e. A similar picture was obtained for 67 out of 69 analyzed variational curves of different bottom communities, both intertidal and subtidal, from soft and hard bottoms, studied by different investigators using different methods.

Figure 85b gives the results of an examination of a community of the digging bivalve molluscs and polychaetes *Macoma baltica* + *Arenicola marina* + *Cardium edule* from the lower intertidal zone of Kola Bay of the Barents Sea described by Zenkevich et al. (1948). The substrate was sand, the total biomass of the animals was 344 g/m^2, the number of species calculated was 17.

It is apparent from the figure that the biomasses of species of this community fall into four groups with distinct breaks in the regions of calculated critical levels.

Figure 86a was taken from the work by Batytskaya (1984) who applied the proposed method to analyze the state of an ecosystem of the Sakhalin bottom fishes. The figure presents data on the catch (biomass) composition of bottom fishes on the shelf of the north-eastern coast of Sakhalin Island. The data include the results of processing of tens of thousand tons of fishes caught. Fifty seven species with the biomass over 0.001% of the total value were plotted on the variational curve. The species fell into four groups (according to biomass) separated by three calculated levels.

Batytskaya also analyzed an ecosystem of bottom fishes from Terpenya Bay which suffered, in the opinion of some ichthyologists, from overexploitation. Here also a structure with four distinctly pronounced allometric ranges was apparent, but the first range comprised only 4 species instead of 11 found about the north-eastern coast of Sakhalin. Navaga, the species predominant in both cases, made 50% of the total catch in Terpenya Bay, and three successive species had a much lower biomass. Batytskaya is inclined to regard this fact as an evidence of some trouble in the fish ecosystem of Terpenya Bay due to its overexploitation. Verification of her supposition on different ecosystems would be of great interest, since its corroboration may yield a method for rapid estimation of the state of a community.

The analysis of the ecosystem structure was made also on the fouling material of experimental plates obtained during 4 years by Oshurkov in Kandalaksha Bay of the White Sea. The results show that the species structure of ecosystems is invariably detected (more than 30 variational curves) at all depths and for all exposure periods (Zhirmunsky and Oshurkov 1984). Figure 86b presents data on biomass of species of a community which was formed at a depth of 5 m for 40 months of exposure. Such a picture was observed in September after the mussel *Mytilus edulis* had increased its biomass ten times (up to 16.8 kg/m^2). It had monopolized the first range, supplanting a few species among which was the bivalve mollusc *Hyatella arctica* that dominated during the second year of succession (ecosystem development).

Of special interest would be a study of successions in relation to competition between species and the dependence of their development on seasonal fluctuations. It should be mentioned that despite changes in the composition of a community and in the biomass of constituent species the communities adapt their structure to the new environmental conditions and optimize species composition and the quantitative relations of species (Nesis 1977). This, for instance, is illustrated by the fact that the community structure, whatever the conditions, comprise groups differing in allometric parameters.

7 General Problems of the Study of Critical Levels in the Processes of Development of Systems

In previous sections, we dealt with various data on the development and organization of biological and other natural systems of different hierarchical levels of integration in terms of manifestation of hierarchy of the critical constants we obtained. These correlations appeared to be manifest in both temporal and spatial characteristics of natural systems of different complexity.

In several cases, the results of analysis led to conclusions that may be of theoretical and practical interest. For example, it has become possible to provide a quantitative basis relating to critical size boundaries between separate systems for a scheme of hierarchically subordinate levels of integration of animate systems distinguished by their functional significance.

A quantitative analysis of data on the abundance of natural and experimental populations allowed us to identify critical levels passing through which may result in rapid growth or disintegration of a population system.

The forms of manifestation of the regularity found are of various characters. In one case, it may be a change in the slope of the curve on a double logarithmic graph at the crossing of the critical boundary, in the other case, it may be an abrupt shift of the curve or a gap between successive values of the series members. In the third case, to reveal changes in the field of the critical range one has to change the kind of dependence or study the change of another index.

It may also be that the members of a group of related phenomena do not (or mostly do not) go beyond the boundaries of one or two adjacent critical ranges.

The cyclicity of development, found empirically by many generations of researchers, where the long evolutionary phase alternates with the shorter revolutionary one (reorganization) is quantitatively represented in natural phenomena by a "unit of development". An analysis of this unit of development shows that the reorganization phase makes for only 20% – 25% of the cycle's duration, which agrees well with results of the analysis of data on the development of the Earth's crust, the phylogenesis and the ontogenesis of organisms.

A comparative analysis of calculated data and boundaries in the Earth's development, identified by geologists on the basis of factual material, was made together with reputable specialists and will certainly make the geochronological scale more precise.

A question of the possibility of enlarging the scope of the regularity found to cover other natural and experimental phenomena is of considerable interest.

A list of the above-mentioned phenomena can be further extended. When studying the literature on populations, we came across an example that could be referred to the species level. Prosser (1973) presents data on the oxygen uptake by

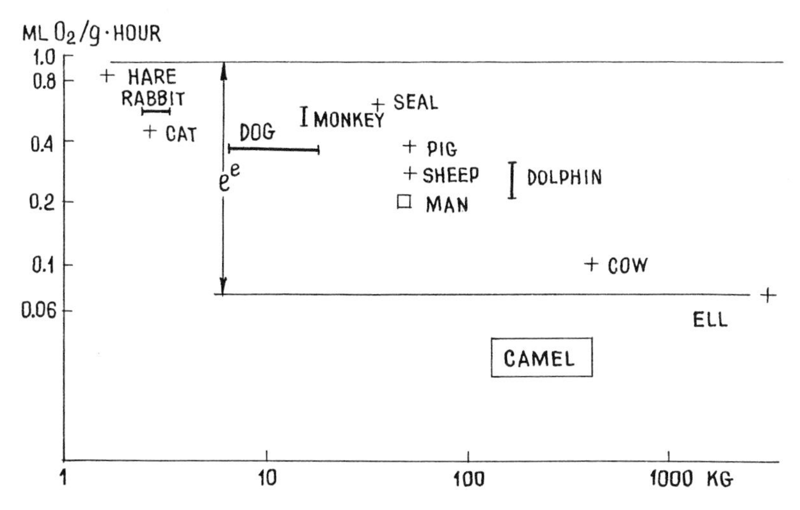

Fig. 87. Oxygen consumption by large mammals, depending on mass. *Abscissa* mass; *Ordinate* oxygen consumption. Logarithmic scale for both axes (Data from Prosser 1973)

large mammals in relation to body weight (Fig. 87). The figure indicates that in 12 out of 13 species the oxygen uptake varies within the range of e^e.

The organization of communities of different levels is of particular interest. This problem has been poorly studied. In fact, biologists usually confine the scope of investigations to the quantitative and qualitative structure of a community, distinguishing dominant, characteristic, and accessory species (Zenkevich 1951).

An analysis of the first three examples from biocenology showed that allometric regularities occur also here. Thus, Korenberg (1979) reports the results of the study of tick encephalitis foci that illustrate the quantitative characteristics of the relationships between several closely associated species (agents – carriers – hosts). On the basis of long-standing field observations, the spatial structure of the distribution area of the agent of tick encephalitis and the structure of the main species-carriers (Ixodoidea) were studied in detail. The basic structural units of this system were identified as: a natural focus, a group of foci, a class of foci, a region of foci, a group of focus regions, and a distribution area of the agent species. The basic elementary unit of the area division is a natural focus. The group of natural foci is composed of several (from 2 to 10–12) island and/or adjacent natural foci. A group of foci is a territorial complex consisting of foci of the same type or close in numbers and intensity of virus circulation. A class of foci comprises up to several dozens of groups of foci close in type. The region of foci is, as a rule, a unique combination of several classes of foci that in every case determines its epizootic structure and epidemic peculiarity. Observations revealed 69 regions of foci united into seven groups. The distribution area of the virus consists of 20000–30000 natural foci. Each focus functions as an independent biocenotic system, which is reflected in many years dynamics of infection within its area.

Based on concrete data on the composition of these structural units, we obtain the following series of values: 1–12, 210–250, 3000–4000, ≥20000–30000

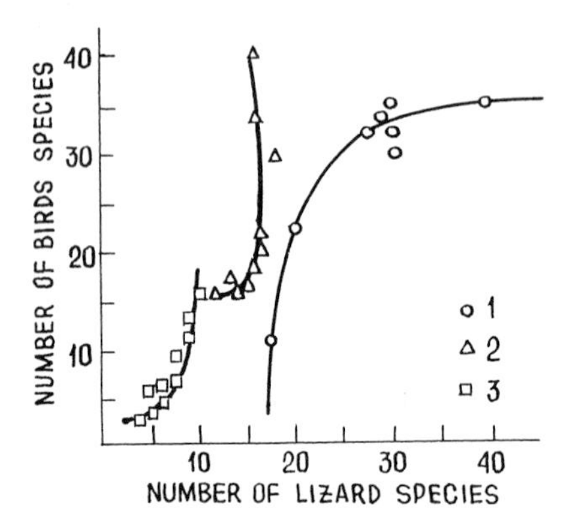

Fig. 88. A comparison of species diversity of birds and lizards on three continents; *1* from Australia; *2* from Africa (Kalahari desert); *3* from North America (Data from Wittacker 1975, p. 117)

foci. The first group of values in this series corresponds to the initial range of changes in critical numbers, the next two agree well with the above-mentioned (Sect. 6.1) critical levels, and the last one is close in the order of values to the fourth critical level.

Figure 88 gives a comparison between species diversity of birds and lizards based on data of selective descriptions of desert regions of three continents (Whittaker 1975). As follows from the figure, the critical number of species – maximal and minimal – is about 15 for both birds and lizards.

In Sect. 6.4, we considered marine ecosystems of bottom invertebrates, and fishes and fouling communities. In all cases, the curves of biomass distribution for the constituent species have definite critical values of biomass at which the character of the curve changes. These critical values constitute a geometric progression with a module e^e and divide subsystems that, taken as a whole, form the ecosystem structure.

As pointed out in Sect. 3.2, the allometric correlation was introduced into biology by Huxley (1932) and Schmalhausen (1935). Besides biology, it has been applied in various branches of science for analyzing physical, chemical, economical and social phenomena (Needham 1950; Von Bertalanffy 1968; Kuzmin et al. 1972; Savageau 1979).

However, none of the above workers have considered the alteration of evolutionary and revolutionary periods in development, the correlation between them, and the relationships between the consecutive cycles of development, as done in the present study.

Naturally, a question arises as to the possibility of applying the critical correlation to the phenomena of inorganic nature. Apart from the Earth's development, we have studied some other such phenomena (Kuzmin and Zhirmunsky 1980a, b, c; Zhirmunsky and Kuzmin 1986).

Let us consider a classic example from hydrodynamics. When gasses or liquids flow through tubes or the circulatory system there is a change from laminar current to a turbulent one. This process is described by the equation: $\lambda = A/Re^m$, where λ is a coefficient of friction against the wall, Re is the Reynolds number, and A and m are parameters true for limited ranges of changes in Re numbers (Fabrikant 1964). The Re number, which characterizes the change from laminar current to a turbulent one corresponding to the fading of the whirls when the liq-

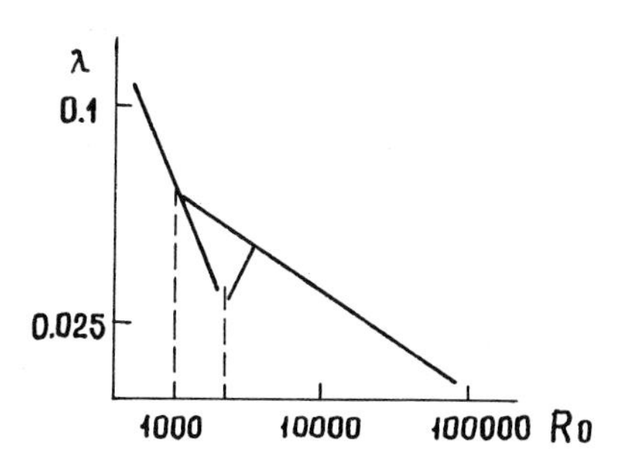

Fig. 89. Dependence of coefficient of wall friction on Reynolds numbers when liquid flows in a tube. *Abscissa* Reynolds numbers (*Re*); *Ordinate* coefficient of wall friction (λ). Logarithmic scale for both axes (Data from Fabrikant 1964)

uid enters the tube, was determined experimentally. In the field of numbers exceeding this value, the least stimulus is sufficient for the current to abruptly change from laminar to turbulent (Fig. 89).

For laminar currents, the Puazeil's formula $\lambda = 64/\mathrm{Re}$ is true (Fabrikant 1964). The strong influence of wall roughness upon the transition to turbulent current and a leap-like character of the transition from laminar to turbulent current enable us to assume a possible retardation in the effect of the Reynolds number on the value of the wall friction coefficient.

The formula for the friction coefficient is as follows: $\lambda = (P_1 - P_2)/[\varrho(v^2/2)(L/d)]$, where P_1, P_2 are pressures in two adjacent units, ϱ the density of the liquid, v the velocity of liquid flow, L, d the length and diameter of the tube.

We assume that the initial value of the friction coefficient is that which corresponds to the loss of all kinetic energy for friction, i.e., $P_1 - P_2 = \varrho(v^2/2)(L/d)$. From this follows $\lambda = 1$. Hence, according to Puazeil's formula, $\mathrm{Re}_* = 64/\lambda \cdot e^e = 970$. Experimental data for round cylindrical tube give $\mathrm{Re}_{\mathrm{round}} = 1\,000-1\,300$ (Sedov 1977), i.e., the value close to the calculated one (without taking into account re-regulation during the experiment).

Calculated and actually observed Reynolds numbers (*Hydrodynamics of blood circulation* 1971) for the blood circulation system of dog are given in Table 33. As follows from data given in the table, the calculated value for Reynolds numbers obtained by dividing consecutive values of Reynolds numbers by e^e,

Table 33. Calculated and experimentally obtained values of Reynolds numbers for the blood circulatory system of dog (From *Hydrodynamics of blood circulation* 1971)

System	Vessels	$\mathrm{Re}_{\mathrm{calc}}$	$\mathrm{Re}_{\mathrm{exp}}$
Arterial	Aorta	1670[a]	1670
	Big arteries	110	130
	Principal arterial branches	–	27
	Terminal arteries	7.8	12
Venous	Terminal veins	6	6.5
	Principal vein branches	–	12
	Big veins	91	72
	Vena cava	1375[a]	1375

[a] Numbers taken as the beginning of counts

beginning with values for aorta and vena cava, appear to be close to the corresponding values in the series obtained experimentally: aorta – big arteries – terminal arteries; vena cava – big veins – terminal veins.

This evidence and the data discussed earlier (Zhirmunsky et al. 1980; Kuzmin and Zhirmunsky 1980b) encourage us to believe that the critical correlation e^e is widely spread also in inorganic nature.

When discussing the results of the present study connected with the regularity of correlation between the characteristics of systems at critical points, we were often asked: What is the reason for such a regular manifestation of correlation e^e?

No less remarkable is the fact of the extremely frequent occurrence of power functions when processing experimental data. This is reflected, on the one hand, in the theory of size and, on the other, in a large-scale production of special technical papers for processing data with a double logarithmic scale.

The presence of critical points in allometric development is well realized and, for a great number of concrete systems, experimentally studied. This leads to the conclusion that methods for processing data by power functions need to be supplemented by restrictions concerning the limits of application of these functions as models of real processes. This requires expanding the space of system parameters which would make it possible to detect the position of a critical point.

From the analysis of properties of developing systems, we conclude that relaxations of characteristics of these systems are universal and have a decisive influence on the formation of critical points beyond which the system appears in the area of instability. The constant correlation between the consecutive critical values of an argument – e^e – for an allometric relaxation system distinguishes a section of power function where development is steady.

Since real systems in which relaxation is in principle absent are unknown, we suppose that for processes described by power function as a model the correlation revealed is as universal as power function when applied to processing data and modeling system dynamics.

Also, a question arises: Why is power function so widely used for describing interconnection between the characteristics of different systems? Under constant internal and external conditions development is known to follow the law of enlarged reproduction (growth of interest on interest), which corresponds to exponential growth. A change of reproduction conditions results in a change of exponential growth rates. At separate stages of development these rates can either increase or decrease. The stages with increasing rates are, as a rule, not so long compared to those at which the rates of exponential growth fall. Such regimes with falling rates are described by a power function which in the given relaxation case is rough.

Consequently, change of the analytical form of model with an increase in the number of its parameters, as compared to power function, but without a changing tendency will lead to no improvement of the approximation level of the available measurements results. We have used this effect for detecting the position of critical points in case they were not clearly represented. In fact, the examples considered confirm that for a rough system different models give nearly equally good results within the framework of one tendency until the critical point is reached, whereas beyond the critical point the results are sharply different. Hence, power function as a rough model with the minimal number of parameters that describes the development of systems at stages characterized by falling growth rates is, really, fairly universal.

The purpose of our study was to establish the boundaries of application of this model. Taking into account the influence of relaxation of the character of development, it was possible to find a proper correlation.

It is in principle clear that any system can function and develop steadily until it attains a certain degree of complexity (or complexity of organization) and growth after which regulatory mechanisms operating in this system fail to fulfill their integrative function and reorganization is required.

However, the correlation revealed is manifested in new and newer examples, and every time, though expected, astonishes us with its wide, apparently universal, scope.

Ideas developed in our works received not only support but also further elaboration in the publications of some authors (Nesis 1982; Polikarpov 1983; Batytskaya 1984; Vorovich et al. 1984; Zaguskin 1984; Konovalov et al. 1984; Efimov et al. 1985).

After having acquainted himself with the main result of our studies – the e^e correlation between consecutive values of the arguments – Polikarpov (1983) "checked the applicability of a dependence between the critical levels described in the monograph for generalizations he had made in the domain of radiobiological structure over a whole known range of dose rates of ionizing radiation" and drew the conclusion that "the structure presented, resting on generalizations of all data on biological chronic effects of ionizing radiation, is well reproduced by means of the universal correlation e^e deduced by A.V. Zhirmunsky and V. I. Kuzmin. Obviously, it is not a mere formal coincidence but rather an indication of the existence of critical structures and processes responsible for the resistance to ionizing radiation" (p. 569).

Zaguskin (1984) worked out an evolutionary classification of the integration levels of biorhythms. "Using the methodology of A.V. Zhirmunsky and V.I. Kuzmin in reference to such parameters of the biosystem as the average and maximum densities of energy flow, it was possible to prove the validity of discrimination of the main levels of biological integration – cell, organism, biocenosis, biosphere, and two intermediate levels between them, formed, accordingly, via integration of homogeneous and heterogeneous elements. Each level is found to be coincident with the appearance of new integral biorhythms of functional changes (from 290–580 μs, 43–86 ms, etc. to 20000–40000 years), functional regulation (from 1–2 s, 15–30 s, 4–8 min, etc. to 4.6–9.3 million years), structural changes (from 66–132 ms, 1–2 s, etc. to 70–140 million years) and structural regulation (from 1–2 h, 15–29 h, 9–18 days, 4.5–9 months, etc., to 1–2 billion years). Experimental studies into the hierarchy of biorhythmical periods of the cell and its microstructures show that a period of a corresponding biorhythm increases (no more than by one order) when the stability is maintained at the cost of an increase in effectiveness under conditions of restricted energy reserve. The period decreases when the stability is maintained at the cost of active change, provided the energy reserve is sufficient. Thus, the current state and stability of the cell and its microstructures could be estimated by deviation in the values for periods of biorhythms. A great deal of literature evidence support the numerical values of biorhythmical periods in the hierarchy of all levels of biosystems. In conclusion, the constructive approach described in the book by A.V. Zhirmunsky and V. I. Kuzmin should be specially emphasized" (Vorovich et al. 1984, pp. 96–97).

In the course of discussing our book new trends of research were outlined. The authors of an imitative model of the Azov Sea ecosystems, who were awarded the State Prize for 1983, believe that their work plus our book "might probably complement each other on the elaboration of general criteria of quantitative assessment and forecast of the stability of all biosystems" (Vorovich et al. 1984, pp. 96).

Konovalov et al. (1984) think it necessary to regard, in the light of our findings, the dynamics of social-economic systems for scientifically grounded prediction of possible changes.

A number of assessments obtained with the aid of critical constants was corroborated by research where our concepts were not applied, for instance in calculations of a potential human life span of 167 years (Zhirmunsky and Kuzmin 1980). Prokofiev (1983) came to the conclusion, on the strength of starvation data, that in man a weight loss of 40% leads to fatal termination, while one of 30% is the maximum permissible. By extrapolating data on weight loss to age, down to the limiting level, he determined a maximum permissible life time of men and women in the last century in Belgium as being 160 years.

During an investigation of regularities in the change of nuclear asymmetry of elements in the periodic table (Evtikhiev et al. 1984) it was found that local minima of dependence between the asymmetry and the number of protons in the isotope nucleus correspond to the critical levels of the unit of development.

It is shown that the position of vertical layers in the atmosphere and the ocean is in the line with the main boundaries of the unit of development (Kuzmin 1984, pp. 164–173).

When analyzing the length of Phanerozoic periods Efimov et al. (1985) considered the boundaries of geochronological scales taken from 17 different sources, including those we had calculated, by modeling critical boundaries with the use of the unit of development. In this case, the remaining 16 sources rely upon empirically gained data that fit well the overwhelming majority of our calculated boundaries. Further, the authors, by employing methods we proposed for determining the age of a synchronized boundary from the duration of an even cycle of development, detect "the main boundary of life development on the earth," which is coincident with the beginning of the wend period.

Our ideas, however, encountered not only support but also sharp criticism (Volkenstein et al. 1983). Criticism is a motive force of development. Therefore, we tried to comprehend the keynote of the criticism so that we should take it into account in future research. To do this required an analysis of fundamental principles, which serve as a basis for a fierce nonacceptance of the whole subject matter of the book. Let us list here some fragments of this analysis.

Volkenstein et al. (1983) write that "the authors expand the equation in a Taylor series, cutting it off after the first two terms. The operation is ungrounded because an equation with deviation argument cannot be treated in such a way, and in the subsequent pages all formulae are erroneous" (p. 570).

Uneasy after such a categorial assertion, we referred to the basic monograph on the Elsgolts and Norkin theory of retarded differential equations (1971) and read on pp. 216–217: "In applied works for an approximate solution ... of retarded differential equations with a small delay a method in power series is widely used. The method is that x $(t-\tau)$ is substituted by several terms of its Taylor expansion in the vicinity of the point

$$x(t-t) = x(t) - \tau\dot{x}(t) + \frac{\tau^2}{2!}\ddot{x}(t) + \ldots + \frac{\tau^m}{m!}x^{(m)}(t) .$$

Such a transition at small τ is allowable only at $m = 1$ since in this case, generally speaking, no equation with a small coefficient in the senior terms is found to appear, and the method applied yields good results." Detailed theoretical grounds for such a possibility were considered by Elsgolts (1955) on pp. 205–211.

The reviewers found as "enigmatic" and "being in compliance with no reasonable model of a developing system" the introduction of an equation with delay argument in the form

$$\frac{dy}{dx} = \lambda e^{-\lambda\tilde{x}}\, y(x-\tilde{x}) \quad \text{(p. 40)} .$$

But in our book we considered (p. 39) a model by Medawar (1945), the Nobel Prize winner, summarizing all known growth curves (size-age correlation), and also the Pearson equation whose integrals define the functions of distribution density in mathematical statistics, and is the most frequently used for treatment of experimental data.

As we have shown, the criticized equation contains both the Medawar and Pearson models as specific cases.

As we recently established (Kuzmin and Zhirmunsky 1986), this equation comprises, as a specific case, also hypergeometric equation, which generalizes the basic equations of the oscillation theory, a measure of the quantity of information, according to Shannon, and is closely allied to the main results of the catastrophe theory.

The analysis of experimental data was most fiercely attacked. Here the reviewers found it possible to negate, without adducing any proof, even obvious things in cases where the text presents explicit information ruling out any dubious interpretation. It cannot be so and that's flat! When there is no alternative, as was the case with the classification of human organismal structures by size range, it is stated: "the correlation is more than dubious and quite senseless" (p. 571).

A detailed analysis of every point of this review (Zhirmunsky and Kuzmin 1986) showed that emotional negation of all the results of the book, without exception, was hardly based on merely reading the book.

The second part of the review (Volkenstein et al. 1963) is reduced to a denial of even a possibility of the existence of common regularities for the animate and inanimate nature. "If one thinks the book over, a conclusion may be made that man's growth, a population of lizards, 'boundaries' of the Quaternary period, of the solar system, and the diameters of canine blood vessels submit to one 'common' law of development" (p. 571).

Exactly! It is a great pity that Professor Volkenstein and his co-authors are not acquainted with the principles of dialectics, the theory of systems and cybernetics, as a science of general laws of reception, retention, transmission and processing of information and links in nature, living organisms, society, and technology.

Konovalov et al. (1984) note: "It might be hard for some of the readers and probably even for prominent scientists of a narrow field to apprehend these regularities as being inherent in the most diverse systems, from molecules, cells and organisms to planets. Nevertheless, such regularities have long been the subject of investigation not only by philosophers but also by mathematicians" (p. 89).

When our concept was discussed some doubts were cast upon the "unique magic number" e^e. It has been primarily found that there is a whole hierarchy of critical constants related to the e number – i.e., the base of natural logarithms (Kuzmin and Zhirmunsky 1986).

Due to investigations carried out in this direction in the last decade, it turns out that we are concerned here with quantitative regularities which are manifested most greatly in the properties of all developing systems.

We register with satisfaction the interest in problems we have raised and hope that the detection of a hierarchy of critical constants will widen the scope of investigations of quantitative relationships of natural systems. If some of the readers will check data at their disposal on the characteristics of natural systems from the viewpoint of our concept, the authors will have attained their goal.

References

Adler YP, Malkova EV, Granovsky YV (1976) Planning experiment in quest of optimal conditions. Nauka, Moscow

Alexandrov VY (1952) On relations between temperature tolerance of protoplasm and temperature environmental conditions, Dokl Acad Nauk SSSR 83(1):149–152

Alexandrov VY (1975) Cells, macromolecules and temperature. Nauka, Leningrad

Allee WC, Emerson AE, Park O, Park T, Schmidt K (1950) Principles of animal ecology. Sounders, Philadelphia-London

Altukhov YP (1974) Population genetics of fishes, Pishchevaya promyshlennost, Moscow

Altukhov YP, Pudovkin AI, Salmenkova EA (1975) Stability in the distribution of frequencies of genes of lactate dehydrogenase and phosphoglucomutase in the system of subpopulation of local schools of fishes as illustrated by *Oncorhynchus nerka* (Walb.). II. Genetika 11(4):54–61

Alven H, Arrenius G (1979) Evolution of the solar system. Mir, Moscow

Anderson FK, Dunn LC, Beasley AB (1964) Introduction of a lethal allele into a feral house mouse population. Am Nat 98(898):57–64

Andronov AA, Pontryagin LS (1937) Rough systems. Dokl Acad Nauk SSSR 14(5):247–250

Andronov AA, Montovich EP, Gordon NI et al. (1967) The theory of bifurcation of dynamic systems on the plane. Nauka, Moscow

Anokhin PK (1962) The anticipating reflection of reality. Vopr Philos 7:97–111

Arakawa M (1954) Fugjiwhara on five centuries of freezing data of lake Suwa in central Japan. Arch Meteor Geoph Biol Ser B, 6 (1–2)

Archimedes (1962) Works. Fizmatgiz, Moscow

Ashby WR (1962) The construction of the brain. Izdatelstvo Inostr Liter, Moscow

Astaurov BL, Gaisinovich AE, Timofeev-Resovsky NV (1970) Biology. In: The Great Soviet Encyclopedia, 3rd edn, v 3:347–356

Astronomichesky Ezhegodnik (1980) Nauka, Moscow

Avtandilov GG (1973) Morphometry in pathology. Meditsina, Moscow

Axelrod DJ (1974) Revolutions in the planet world. Geophytology 4(1/6)

Baltensweiler W (1964) *Zeiraphera griseana* Hübner Leipidoptera: Tortricidae in European Alps: A contribution to the problem of cycles. Can Entomol 96(5):792–800

Baranov AS, Yablokov AV, Valetsky AV (1976) Numbers and its dynamics. In: The sand lizard. Nauka, Moscow: 302–321

Baranov AS, Rozanov AS, Turutina LV (1977) An attempt of determining the numbers of specimens in natural groups of the sand lizard. In: Problems of herpetology. Zool Inst Acad Nauk SSSR, Leningrad 4:23–24

Barott HG (1937) Effect of temperature, humidity and other factors on hatch of hen's eggs and on energy metabolism of chick embryos. Tech Bull US Dep Agr, N 553

Bates M (1945) Observations on climate and seasonal distribution of mosquitoes in Eastern Colombia. J Anim Ecol 14:17–25

Batytskaya LV (1984) Changes in the composition of communities of bottom fishes on the shelf of eastern Sakhalin due to intensive fishery. Sov J Mar Biol (Engl Transl Biol Morya) 10(2):98–104

Bauer ES (1935) Theoretical biology. Vses Inst Exp Med, Moscow Leningrad

Bautin NN, Leontovich EA (1976) Methods and ways of qualitative investigation of dynamic systems on the plane. Nauka, Moscow

Bayley N (1970) The mathematical approach to biology and medicine. Mir, Moscow

Bekesy G, Rosenblith WA (1958) The mechanical properties of the ear. In: Stevens SS (ed) Handbook of experimental psychology. Wiley, New York: 1075–1115
Beklemishev VN (1964) On general principles of organization of life. Bull Mosk Obshch Isp Prir Otd Biol 69(2):22–38
Belousov VV (1975) Principles of geotectonics. Nedra, Moscow
Belousov VV (1978) Endogenous regimes of the continents. Nedra, Moscow
Berggren WA, Van Couvering JA (1974) Late Neogene: biostratigraphy, geochronology and palaeoclimatology of the last 15 million years in marine and continental sequences. Palaeogeogr Palaeoclimatol Palaeoecol 16:1–216
Bernal J (1969) The origin of life. Mir, Moscow
Bernar K (1878) A handbook of general physiology. St Petersburg
Berry BJ (1968) The ecology of an island population of the house mouse. J Anim Ecol 37(2):445–470
Bertalanffy L von (1949) Das biologische Weltbild. Francke, Bern
Bertalanffy L von (1952) Problems of life. Watts, London
Bertalanffy L von (1960) Principles and theory of growth. Fundamental aspects of normal and malignant. Elsevier, Amsterdam
Bertalanffy L von (1968) General system theory. Foundations, development, applications. Braziller, New York
Berzin AA, Yablokov AV (1978) Numbers and population structure of main exploited species of cetaceans in the World Ocean. Zool Zh 57(12):1771–1785
Biebl R (1965) Cytological bases of plant ecology. Mir, Moscow
Biology Data Book (1964) Altman PL, Dittmer DS (eds). Fed Am Soc Exp Biol, Washington DC
Birney EC, Grant WE, Baird DD (1976) Importance of vegetative cover to cycles of *Microtus* populations. Ecology 57(5):1043–1051
Blinkov SM, Glezer II (1964) The human brain in figures and tables. Meditsina, Leningrad
Bogdanov AA (1912) The universal organizational science (tectology), part I. St Petersburg
Bogdanov AA (1925–1929) The universal organizational science (tectology), parts 1–3. Kniga, Leningrad Moscow
Bogomolets AA (1940) Prolongation of life. Izdatelstvo Acad Nauk USSR, Kiev
Brody S (1927) Growth and development. Univ Missouri Agr Exp Sta
Brody S (1945) Bioenergetics and growth. Hafner, New York
Brotskaya VA, Zenkevich LA (1939) Quantitative registration of bottom fauna of Barents Sea. Tr VNIRO 4:5–126
Brown HC (1926) The materialist's view on the concepts of levels. J Philos 23(5)
Brown AA, Rakhishev AP (1975) Cells and tissues of a living organism. Kazakhstan, Alma-Ata
Budyko MI (1977) The global ecology. Mysl, Moscow
Bullock TH (1955) Compensation for temperature in the metabolism and activity of poikilotherms. Biol Rev Camb Philos Soc 30(3)
Chetverikov SS (1905) Waves of life (from lepidopterological observations during summer of 1903). Dnevnik zool otd Ob-va Lyubitelei Yestestvoznaniya, Antropologii, Etnographii 3(6):106–110
Child CM (1929) The physiological gradients. Sammelreferat. Protoplasma 5(3):447–476
Childs W (1962) Physical constants. Fizmatgiz, Moscow
Chizhevsky AL (1976) Earth's echo of the solar storms, 2nd edn. Mysl, Moscow
Chuev YV, Mikhailov YB, Kuzmin VI (1975) The forecasting of quantitative characteristics of processes. Sov Radio, Moscow
Comfort A (1967) Biology of senility. Mir, Moscow
Commission on Fur Seals of the North Pacific Ocean (1971) Report on researches (1964–1965). Head-quarters of Commission, Washington
Crawford MK, Genzel R, Harris AI, Jaffe DT, Lacy JH (1985) Mass distribution in the galactic centre. Nature 315(6019):467–470
Cuvier J (1937) Discussions about revolutions on the surface of the globe and changes caused by them in the animal kingdom. Biomedgiz, Moscow Leningrad
Dajoz R (1972) Précis d'Ecologie, 2nd edn. Dunod ed., Paris
Darwin C (1859) The origin of species by means of natural selection or the preservation of favoured races in the struggle for life. Murrey, London
David H (1977) Quantitative ultrastructural data of animal and human cells. Thieme, Leipzig

Davidovsky IV (1958) The pathologic anatomy and pathogenesis of human diseases, vol 2. Medgiz, Moscow
Davidson J, Andrewartha HG (1948) The influence of rainfall, evaporation and atmospheric temperature on fluctuations in the size of natural population of *Thrips imaginis* (Thysanoptera). J Anim Ecol 17(2):193–222
Day WA (1974) Thermodynamics of simple systems with memory. Izdatelstvo Inostr Liter, Moscow
Dewcar E (1978) Cellular interactions in animal development. Mir, Moscow
Dice LR (1952) Natural communities. Univ Michigan Press, Ann Arbor
Dobzhansky Th (1937) Genetics and the origin of species. Columbia Univ Press, New York
Eliseev VG (1972) Histology. Meditsina, Moscow
Efimov AA, Zakoldayev YA, Shpitalnaya AA (1985) Astronomic bases of absolute geochronology. In: Sundial and calendar systems of the peoples of the USSR. Glavnaya Astron Observ, pp 185–201
Elsgolts LE (1955) Quantitative methods in mathematical analysis. Gos Izdatelstvo Tekhn-Teor Liter, Moscow
Elsgolts LE, Norkin SB (1971) An introduction to the theory of equations with a deviating argument. Nauka, Moscow
Engelgart VA (1970) Integratism as a way from the simple to the complex in apprehending life phenomena. Nauka, Moscow
Engels F (1878) Herrn Eugen Dühring's Umwälzung der Wissenschaft. Philosophie, politische Ökonomie, Sozialismus. Leipzig
Eskov EK (1979) Acoustic signalling of social insects. Nauka, Moscow
Evtikhiev NN, Kuzmin VI (1982) Cybernetics of systems development. Moscow Inst of Radioengineering, Electronics and Automation, Moscow
Evtikhiev NN, Kuzmin VI, Grakin AI, Lukyanova VG (1984) A study of the nuclear asymmetry changes regularities of the DI Mendeleev table elements. In: Problems of cybernetics. Arrangements and systems. Moscow Inst of Radioengineering, Electronics and Automation, Moscow: 3–9
Fabrikant NY (1964) Aerodynamics. Nauka, Moscow
Feigenbaum MJ (1980) Universal behavior in nonlinear systems. Los Alamos Sci 1(1):4–27
Finean J (1970) Biological ultrastructures. Mir, Moscow
Fisher F (1968) The nature of a critical condition. Mir, Moscow
Fleishman BS (1971) Elements of theory of potential efficiency of complex systems. Sov Radio, Moscow
Flint VE (1977) The spatial structure of populations of small mammals. Nauka, Moscow
Ford JJ (1937) Fluctuations in natural populations of Collembola and Acarina. Anim Ecol 6(1):98–111
Fox LR (1975) Some demographic consequences of food shortage for the predator, *Notonecta hoffmanni.* Ecology 56(4):868–880
Frisch K von (1977) Aus dem Leben der Bienen, 9th edn. Springer, Berlin Heidelberg New York
Frolkis VV (1969) The nature of senility. Nauka, Moscow
Gause GF (1934) The struggle for existence. Williams & Wilkins, Baltimore
Gelfand SA (1981) Hearing. An introduction to psychological and physiological acoustics. Dekker, New York
General problems of division of the Precambrian of the USSR (1979). Nauka, Leningrad
Gilmor R (1984) Applied catastrophe theory. Mir, Moscow
Gilyarov MS (1954) Species, population and biocenosis. Zool Zh 33(4):769–778
Gilyarov MS (1973) Some basic regulations of ecology of terrestrial invertebrates. Zh Obshch Biol 34(6):795–807
Ginkin GG (1962) Logarithms, decibels, decilogs. Energoizdat, Moscow Leningrad
Gnezdilova SM, Kuzmin VI, Lenskaya GS (1976) Mathematical description of growth processes in sea urchin eggs, nuclei and nucleoli. Sov J Mar Biol (Engl Transl Biol Morya) 2(2):103–106
Golenkin MI (1959) Winners in the struggle for existence. A study of reasons of the conquest of the Earth by angiospermous plants in the middle Crustaceous period, 3rd edn. Uchpedgiz, Moscow
Goodwin BC (1966) Temporal organizations in cells. Mir, Moscow
Grant V (1977) Organismic evolution. Freeman, San Francisco
Gulyaev YV, Godik EE (1984) Physical fields of biological objects. In: Cybernetics of alive. Biology and information. Nauka, Moscow: 111–116
Guretsky K (1974) Analysis and synthesis of control systems with delay. Mashinostroyeniye, Moscow
Gurvich AG (1977) Selected works. The theoretical and experimental researches. Meditsina, Moscow

Hegel GWF (1986) Dissertatio Philosophica de Orbis Planetarum. Philosophische Erörterung über die Planetenbahnen. Verlag Chemie, Weinheim
Henderson IF (1979) Henderson's dictionary of biological terms, 9th edn. Longman, London
Hruban Z, Rechcigl M (1972) Microbodies and related particles. Mir, Moscow
Hufeland CW (1796) Die Kunst, das menschliche Leben zu verlängern. Berlin
Huxley JS (1932) Problems of relative growth. Methuen, London
Huxley TH (1883) Critiques and addresses. Macmillan, London
Hydrodynamics of blood circulation (1971). Mir, Moscow
Ivanova EA (1955) To the question about the relationship between evolutionary stages of the organic world and those of the Earth's crust. Dokl Acad Nauk SSSR 105(7):154–157
Kalabushkin BA (1976) The intrapopulation changeability in the recent and middle Holocene *Littorina squalida*. Zh Obshch Biol 37(3):369–377
Karogodin YN (1975) The atlas of lithologo-paleogeographical maps of the USSR. In: Paleogeography of the USSR, vol 4. Nedra, Moscow, p 175
Karogodin YN (1980) The sedimentational cyclicity. Nedra, Moscow
Kaufman U (1982) The planets and the moons. Mir, Moscow
Khain VE (1973) General geotectonics. Nedra, Moscow
Khesin YE (1967) Sizes of nuclei and functional state of cells. Meditsina, Moscow
Khinchin AY (1961) Chain fractions. Fizmatgiz, Moscow
Klige RK (1980) The ocean level in the geological past. Nauka, Moscow
Kobrinsky HE, Kuzmin VI (1981) Accuracy of economical-mathematical models. Finansy Statistika, Moscow
Konovalov SM (1971) A differentiation of local stocks of the blueback salmon *Oncorhynchus nerka* (Walb.). Nauka, Leningrad
Konovalov SM (1972) The structure of the isolate of the blueback salmon *Oncorhynchus nerka* (Walb.) from the Azabachye Lake. Zh Obshch Biol 3(6):668–682
Konovalov SM (1980) Population biology of Pacific salmon. Nauka, Leningrad
Konovalov SM, Kontrimavichus VL, Krasnov EV (1984) AV Zhirmunsky, VI Kuzmin, Critical levels in the processes of development of biological systems, a review. Ecology 3:88–89
Kontorin VV (1980) The mathematical simulation of the Baikal arctic cisco population. Nauka, Novosibirsk
Koptev GS, Pentin YA (1977) A calculation of molecules' fluctuations. Izdatelstvo Mos Univ, Moscow
Kordonsky KB (1963) The theory of probability in engineering. Fizmatgiz, Moscow Leningrad
Korenberg EI (1979) The biochorological structure of a species on the example of a taiga tick. Nauka, Moscow
Kramer G (1975) Mathematical methods of statistics. Mir, Moscow
Krasilov VA (1973) A step-like character of evolution and its reasons. Zh Obshch Biol 34(2):227–240
Krasnov II (1977) Stratigraphische Korrelation der Quartärablagerungen im östlichen Gebiet der fennoskandischen Verlosungen. Schriftenr Geol Wiss 9:69–79
Krasovsky NN (1959) Some tasks of the theory of movement stability. Gostekhizdat, Moscow
Krasovsky VI, Shklovsky IS (1957) Possible influences of flashes of supernews on evolution of life on the Earth. Dokl Acad Nauk SSSR 116(2):197–199
Krats KO, Khiltova VY, Vrevsky AB et al. (1980) Division into periods of tectonic events of the Precambrian. In: Tectonics of the Early Precambrian. Nauka, Leningrad
Kremyansky VI (1969) Structural levels of the living matter. Nauka, Moscow
Krishtofovich VN (1957) Fossil botany. Gos Nauch-Tekhn Izd-vo Neft Gornoprom Liter, Leningrad
Krylov DG (1975) Tendency to grouping in spatial distribution of small mammals in a forest habitat. Ecol Pol 23(2):335–345
Krylov YG, Yablokov AV (1972) Power polymorphism in the skull structure of the red mouse *Clethrionomys glareolus*. Zool Zh 51(4):576–584
Kuzmin VI (1984) Introduction to informatics. Moscow Inst of Radioengineering, Electronics and Automation, Moscow
Kuzmin VI, Lenskaya GS (1974) The model of stable growth of biological systems. Zh Obshch Biol 35(5):778–791
Kuzmin VI, Zhirmunsky AV (1980a) Models of critical levels in stable allometric development of biological systems. I. Modelling of the critical levels in allometric development. Ontogenesis 11(6):563–570

Kuzmin VI, Zhirmunsky AV (1980b) The law of critical levels of allometric development of systems. Dokl Acad Nauk SSSR 225(6):1513–1516
Kuzmin VI, Zhirmunsky AV (1980c) Critical levels in allometric development of systems. Preprint of the Inst biologii morya, Dalnev Nauch Center, Vladivostok
Kuzmin VI, Zhirmunsky AV (1986) The model of critical levels in development of systems, I and II. Preprint of the Inst biologii morya, Dalnev Nauch Center, Vladivostok
Kuzmin VI, Lebedev BD, Chuev YV (1972) Ways of improvement of analytical patterns of development. In: Problems of cybernetics. Nauka, Moscow, 24:5–14
Lack D (1966) Population studies of birds. Clarendon, Oxford
Lavrenko EM (1965) On levels of studying the organic world in connection with knowledge of vegetational cover. In: Problems of modern botany, vol 2. Nauka, Moscow-Leningrad: 364–378
Leopold A (1943) Deer irruptions. Wiscons Cons Dept Publ 321:3–11
Lichkov BL (1945) Geological periods and evolution of the living matter. Zh Obshch Biol 6(3):157–182
Lindsey P, Norman D (1974) Processing of information of man. Mir, Moscow
Lorents G (1897) Elements of higher mathematics. Tovarishchestvo Sitina, Moscow
Lyell C (1830–1833) Principles of geology 1–3
Ma S (1980) Modern theory of critical phenomena. Mir, Moscow
Macfadyen A (1963) Animal ecology. Aims and methods, 2nd edn. Pitman, London
Magnetostratigraphical scale of Phanerozoic and the regime of geomagnetic field of inversion (1976) In: Geomagnetic studies, vol. 17. Nauka, Moscow, pp 45–52
Makarycheva AM (1983) Resistance of sea urchin *Strongylocentrotus nudus* to high temperatures in embryonic and early larval stages of development. Sov J Mar Biol (Engl Transl Biol Morya) 9(2):79–83
Man: Medico-biological data (1977) Meditsina, Moscow
Markevich AP (1968) Levels of life organization and principles of their study. In: Integration and biology. Naukova Dumka, Kiev: 71–80
Marrel J, Kettle S, Tedder J (1968) The theory of valency. Mir, Moscow
Marriott RA (1964) Stream catalogue of the World River Lake system, Bristol Bay, Alaska. US Fish Wildl Serv, Spec Sci Rept Fish: 494
Marochnik LS, Suchkov AA (1984) The Galaxy. Nauka, Moscow
McCormak B, Seliga Th (eds) (1978) Solar-terrestrial influences on weather and climate. Reidel, Dordrecht
Mechnikov II (1964) Essays in optimism. Nauka, Moscow
Medawar PB (1945) Size, shape and age. Essays on growth and form. Univ Press, Oxford
Mendeleev DI (1959) Solutions. Izdatelstvo Acad Nauk SSSR, Moscow
Mendelson LA (1959) Theory and history of economical crises and cycles, vol 2. Sotsekgiz, Moscow
Migdal AB (1975) Qualitative methods in the quantum theory. Nauka, Moscow
Milanovsky EE (1979) To the problem of the origin and development of platform linear structures. Vestn Mosk Univ Ser 4 Geol 6:29–53
Mina MV, Klevezal GA (1976) Animal growth. Nauka, Moscow
Morozov VP (1977) Biophysical basis of vocal speech. Nauka, Leningrad
Nalivkin DV (1980) Articles on geology of the USSR. Nedra, Leningrad
Nalivkin VD (1975) Periodicity of transgressions and regressions. In: Paleogeography in the USSR, vol 4. Nedra, Moscow: 162–167
Nalivkin VD, Kuzmin VI, Lukyanova VG (1982) Natural boundaries in a series of oil and gas deposit distribution by size. Dokl Acad Nauk SSSR 266(4):948–951
Nalivkin VD et al. (1984) Discretion in distribution and development of natural objects. In: Tectonics and capacity of oil- and gas-bearing folded belts. Kyrgyzstan, Frunze: 27–35
Nasonov DN (1962) The local reaction of protoplasm and spreading excitation. Izdatelstvo Acad Nauk SSSR, Moscow, Leningrad
Nasonov DN, Alexandrov VY (1940) Reaction of the living matter to external effects. Denaturation theory of damage and stimulation. Izdatelstvo Acad Nauk SSSR, Moscow, Leningrad
Naumov NP (1964) On methodological problems in biology. Nauch Dokl Vyssh Shk Filos Nauki 1:136–145
Needham J (1950) Biochemistry and morphogenesis. Univ Press, Cambridge

Nesis KN (1977) Basic ecological concepts in application to the sea communities. Community as a continuum. In: Biological productivity of the ocean, vol 2. Nauka, Moscow: 5–13
Nesis KN (1982) Universal critical limits: a review. Sci USSR 5:121–123
Nevorotin AI (1977) Bordered bubbles and their functional significance in animal cells. Tsitologiya 19(1):5–14
Newell ND (1967) Revolutions in the history of life. Geol Soc Am Spec Pap (Reg Stud) 89:63–91
Nicholson AJ (1954) An outline of the dynamics of animal populations. Aust J Zool 2(1):9–65
Noll W (1959) The foundations of classical mechanics in the light of recent advances in continuum mechanics. In: The axiomatic method, with special reference to geometry and physics. North-Holland, Amsterdam: 266–281
Norques RM (1977) Population size fluctuation in the evolution of experimental cultures of *Drosophila subobscura.* Evolution 32(1):200–213
Novikoff AB (1945) The concept of integrative levels and biology. Science 101(2618):209–215
Ober-Krie J (1973) Managing an enterprise. Progress, Moscow
Odum EP (1971) Fundamentals of ecology, 3rd edn. Saunders, Philadelphia
Ognev SI (1931) Animals of eastern Europe and northern Asia. Gosizdat, Moscow, Leningrad
Ostroumov AA (1912) The second year of sterlet's growth. Tr Estestvoispyt Kazan Univ 45(1):1–24
Ostroumov AA (1918) The further course of sterlet's growth. Tr Estestvoispyt Kazan Univ 49(6):1–36
Ovchinnikov YA (1980) Main tendencies in development of physicochemical biology. Priroda (Mosc) 2:2–12
Park T, Gregg EV, Lutherman CZ (1941) Studies in population physiology. X. Interspecific competition in populations of granary beetles. Physiol Zool 14:395–430
Patashinsky AZ, Pokrovsky VL (1975) The theory of fluctuation of phase transitions. Nauka, Moscow
Pavlov AP (1924) On some still poorly studied factors of animal extinction. In: Pavlova MV (ed) Causes of animal extinction in last geological epochs. Petrograd, Moscow: 89–130
Pavlov IP (1949) In memory of P Geidengain. In: The complete works. vol 5. Izdatelstvo Acad Nauk SSSR, Moscow, Leningrad, pp 154–163
Pavlov IP (1952a) Papers on various problems of physiology, speeches etc. The complete works. vol 6. Izdatelstvo Acad Nauk SSSR, Moscow, Leningrad
Pavlov IP (1952b) Lectures on physiology, 1912–1913. Izdatelstvo Acad Med Nauk SSSR, Moscow
Pearl R (1924) The curve of population growth. Proc Am Philos Soc 63
Perna N (1925) Life and work rhythms. "Petrograd", Moscow, Leningrad
Piaget J (1968) Psychology of intellect. In: Selected psychological works. Prosveshcheniye, Moscow
Pimental D, Stone FA (1968) Evolution and population ecology of parasite-host systems. Can Entomol 100:655–662
Pimentel G, Spartly R (1970) Chemical bonding clarified through quantum mechanics. Holden Day, San Francisco
Pirson K (1911) The grammar of science. Shipovnik, St Petersburg
Poglazov BF (1970) Assembling of biological structures. Nauka, Moscow
Polikarpov GG (1983) AV Zhirmunsky, VI Kuzmin, Critical levels in the processes development of biological systems, a review. Zh Obshch Biol 44(4):568–569
Polyansky YI (1959) Temperature adaptations in infusoria. Tsitologiya 1(6):714–727
Polyansky YI (1971) On peculiar features of progressive evolution on cellular level. Zh Obshch Biol 32(5):541–548
Popova OA (1971) Biological indices of the pike and the perch in pools of different hydrobiological regime and feed quantity. In: Objective regularities of fish growth and maturation. Nauka, Moscow: 102–152
Poston T, Stuart N (1978) Catastrophe theory and its applications. Pitman, London
Prokofiev EA (1983) Quantitative analysis of growth and forecasting of life length. Abstract of the author's thesis. Institute of Developmental Biology, Moscow
Prosser L (1973) Comparative animal physiology, 3rd edn. Saunders, Philadelphia
Ragozina MN (1961) The development of hen embryo in its correlation with the egg yolk and envelopes. Izdatelstvo Acad Nauk SSSR, Moscow

Ravikovich AI (1969) The development of principal theoretical directions in geology of the 19th century. Nauka, Moscow
Rezanov IA (1980) Great catastrophes in the Earth's history. Nauka, Moscow
Robertis E de, Nowinski WW, Saez FA (1970) Cell biology. Saunders, Philadelphia
Romanoff AI (1967) Biochemistry of the avain embryology. Wiley, New York
Ronov AB (1976) Volcanism, carbonate accumulation, life. Geokhimiya 8:1252–1277
Rosen R (1969) Optimality principles in biology. Mir, Moscow
Routh S, Routh F (1964) Excretion of substances by tadpoles hampering their growth. In: Mechanisms of biological competition. Mir, Moscow: 263–276
Rybakov VA (1984) From the history of ancient Russia. Moscow Univ, Moscow
Sadovsky MA (1983) Size distribution of solid components particles. Dokl Acad Nauk SSSR 269(1):69–72
Sadovsky MA (1984) Structural hierarchy from the speck of dust to the planets. Zemlya Vselennaya 6:4–9
Sadovsky MA, Bolkhovitinov LG, Pisarenko VF (1982) On properties of discrete deposits. Izdatelstvo Acad Nauk SSSR, Fizika Zemli 2
Sadovsky VN (1976) The System. In: Great Soviet Encyclopedia, 3rd edn. 23:463–464
Savageau NA (1979) Allometric morphogenesis of complex systems: Derivation of the basic equations from first principles. Proc Natl Acad Sci USA 76(12):6023–6025
Savelov AA (1960) Plane curves. Fizmatgiz, Moscow
Schindewolf O (1954) Über die möglichen Ursachen der großen erdgeschichtlichen Faunenschnitte. Neues Jahrb Geol Paläontol 10:457–465
Schindewolf O (1963) Neokatastrophysmus? Z Dtsch Geol Ges 114(2):430–445
Schleiden MJ (1846) Grundzüge der wissenschaftlichen Botanik. Leipzig
Schmalhausen II (1929) Analysis of vertebrate growth using the method for the determination of the growth constant. – In: Trudy Fisiko-Matemat. Otdela Vseukrain. Akad. Nauk 12(4)
Schmalhausen II (1935) Determining basic conceptions and method of studying growth. In: The animal growth. Biomedgiz, Moscow: 8–60
Schmalhausen II (1961) Integration of biological systems and their self-regulation. Bull Mosc Obshch Isp Prir Otd Biol 66(2):104–134
Schmalhausen II (1984) Growth and differentiation. Selected works. 1–2 Naukova Dumka, Kiev
Schrödinger E (1945) What is life? The physical aspect of the living cell. Cambridge University Press, Cambridge
Sedov LI (1977) The method of similarity and proportions in mechanics. Nauka, Moscow
Seismic Stratigraphy – applications to hydrocarbon exploration (1977) Ch Payton Mem Am Assoc Petrol Geol, vol 26
Seleshnikov SI (1977) Calendar history and chronology. Nauka, Moscow
Selin NI (1980) Some peculiarities in growth of the mussel *Crenomytilus grayanus* on the artificial substrates in the Wityaz Bay of the Possyet Bay (Sea of Japan). Biol Morya (Vladivost) 3:97–99
Sellars R (1926) The principles and problems of phylosophy. New York
Setrov MI (1966) The significance of the general system theory by L Bertalanffy for biology. In: Philosophical problems of modern biology. Nauka, Leningrad, Moscow: 48–62
Setrov MI (1971) Organization of biosystems. Nauka, Leningrad
Setrov MI (1972) The interaction of basic hierarchical levels of the living matter organization. In: The development of conception of structural levels in biology. Nauka, Moscow: 311–321
Shennon K (1963) Works on theory of information and cybernetics. Izdatelstvo Inostr Liter, Moscow
Shumakov VI, Novoseltsev VN, Sakharov MP et al. (1971) Modelling of physiological systems of organisms. Meditsina, Moscow
Shwarts SS (1969) Evolutionary ecology of animals. Ecological mechanisms of the evolutionary process. Tr Inst Ekol Rast Zhivotn Ural Fil Acad Nauk SSSR 65:1–199
Sigarev VA, Agafonova TK (1975) Some peculiarities of number dynamics of red and common mice in the Saratov region. In: Physiological and population animal ecology, issue 3(5). Izdatelstvo Sarat Univ, Saratov: 134–138
Sims PK (1979) Precambrian subdivided. Geotimes 24(12)
Smirnov VS (1967) Analysis of number dynamics of Arctic fox on the Yamal peninsula and ways of intensification of its hunting, issue 11. Probl Severa: 70–90
Sobolev DN (1924) Outsets of historical biogenetics. Gosizdat Ukraini, Kiev

Sokolov BS (1977) The organic world of the Earth on the way to Phanerozoic differentiation. In: The 250 year anniversary of the USSR Academy of Sciences. Nauka, Moscow: 432–444
Stanek I (1977) Embryology of man. Veda, Bratislava
Stille H (1964) Geotektonische Gliederung der Erdgeschichte. Selected works. Mir, Moscow: 344–394
Streller B (1964) Time, cells and senility. Mir, Moscow
Sukachev VN (1957) General principles and programme of study of forest types. In: Sukachev VN, Zonn SV (eds) Methodic instructions for study of forest types. Izdatelstvo Acad Nauk SSSR, Moscow: 11–104
Sushkin PP (1922) The evolution of terrestrial vertebrates and the role of geological changes. Priroda (Mosc) 3(5):3–32
Svetlov PG (1960) Theory of critical periods of development and its significance in comprehension of the principles of environmental effect on ontogenesis. In: Problems of cytology and general physiology. Izdatelstvo Acad Nauk SSSR, Leningrad Moscow: 263–285
Svetlov PG (1978) Physiology of development. II. Internal and external factors of development. Nauka, Leningrad
Svenson CP, Webster PL (1977) The cell, 4th ed. Prentice-Hall, Englewood Cliffs, New Jersey
Szent-Györgyi A (1960) Introduction to a submolecular biology. Academic Press, London
Tansley AC (1935) The use and abuse of vegetational concepts and terms. Ecology 16(4):284–307
Tarasov VG (1978) Distribution and trophic regionalization of soft bottom communities in Vostok Bay, Sea of Japan. Sov J Mar Biol (Engl Transl Biol Morya) 4(6):889–894
Terao A, Tanaka T (1928) Influence of temperature upon the rate of reproduction in water-flea, *Moina macrocopa* Strauss. Proc Imp Acad Tokyo 4(9):553–555
Terskova MI (1978) A stepwise parabolic equation of bird embryos growth. In: Analysis of growth dynamics of biological objects. Nauka, Moscow: 5–14
The development of conception of structural levels in biology (1972). Nauka, Moscow
The geological structure of the USSR and territory regularities in location of deposits (1968). Nedra, Moscow
The problem of integrity in Modern Biology (1968) Jugai GA (ed). Nauka, Moscow
The problem of levels and systems in scientific cognition (1979). Nauka, Tekhnika, Minsk
The sand lizard (1976). Nauka, Moscow
The Soviet Encyclopedic Dictionary (1980). Sov entsiclopediya, Moscow
Thom R (1968) Dynamic theory of morphogenesis. In: On the way to theoretical biology. Prolegomena. Aldine, Birmingham
Thompson D'Arcy W (1945) On growth and form. MacMillan, New York
Tibbo SW, Graham TR (1963) Biological changes in herring stocks following an epizootic. J Fish Res Board Can 20:435–449
Timofeev-Resovsky NV, Vorontsov NN, Yablokov AV (1977) A short essay on the evolution theory. Nauka, Moscow
Tomashevsky KE (1973) Dynamics of squirrel number in the Upper Volga and adjacent regions. In: Helminths and their hosts. Kalinin: 166–174
Troshin AS, Brown AD, Vakhtin YB et al. (1970) Tsitologiya. Prosveshcheniye, Moscow
Truesdell C (1975) A first course in rational continuum mechanics. Mir, Moscow
Turutina LV (1979) Peculiarities of spatially-genetic intrapopulation structure of higher vertebrates. Author's abstract of candidate's dissertation. Inst Biologii Razvitiya Acad Nauk SSSR, Moscow
Turutina LV, Krylov DG (1979) The population structure heterogeneity of the red mouse. Priroda (Mosc) 5:111
Turutina LV, Podmarev VI (1978) The territorial distribution of phenes and singling out genetically-spatial groups in population of the sand lizard (*Lacerta agilis* L.). In: Physiological and population animal ecology, issue 5(7). Izdatelstvo Sarat Univ, Saratov: 72–77
Tuttle MD (1976) Population ecology of the grey bat (*Myotis grisescens*): Factors influencing growth and survival of new volant young. Ecology 57(3):587–595
Urlanis VI (ed) (1978) Population of the world countries. Statistika, Moscow
Uryson AM (1973) Age dynamics of body sizes of children and teenagers at ages between 4 and 18 years. In: Child growth and development according to anthropological data. Izdatelstvo Mos Univ, Moscow: 4–54

Ushakov BP (1959) Temperature tolerance of tissues as one of the diagnostic specific features in poikilotherms. Zool Zh 38(9):1292–1302
Ushakov BP (1963) On classification of animal and plant adaptations and on the role of cytoecology in working out adaptational problems. In: Problems of animal cytoecology. Izdatelstvo Acad Nauk SSSR, Leningrad, Moscow: 5–20
Valentine JW (1973) Evolutionary paleontology of the marine biosphere. Prentice Hall, Englewood Cliffs, New Jersey
Varley GC (1949) Population changes in German forest pests. J Anim Ecol 18:117–122
Vavilov SI (1943) Isaac Newton. Izdatelstvo Acad Nauk SSSR, Moscow, Leningrad
Vedenev MP, Kremyansky VI, Shatalov AG (1972) A conception of structural levels in biology. In: Development of conception of structural levels in biology. Nauka, Moscow: 7–70
Vernadsky VI (1978) The living matter. Nauka, Moscow
Vinberg GG (ed) (1968) Methods of determining production of aquatic animals. Vysheish Shkola, Kiev
Vinberg GG (ed) (1979) General principles for the study of aquatic ecosystems. Nauka, Leningrad
Virkhov R (1865) Cellular pathology as a doctrine based on physiological and pathological histology. Tip Millera, Moscow
Volkenstein MV, Livshiz MA, Lisov YP (1983) On the law of general development: a review. Zh Obshch Biol 44(4):569–571
Volskis RS (1973) Productivity of a species and its study within the range of the area. Mintis, Vilnyus
Volterra V (1976) The mathematical theory of the battle for survival. Nauka, Moscow
Vorovich II, Gorstko AB, Zaguskin SL (1984) AV Zhirmunsky, VI Kuzmin, Critical levels in the processes of development of biological systems, a review. Izv Sev Kavk Nauchn, Tsentra 4:96–97
Wagner, Suyder DP (eds) (1978) Populations of small mammals under natural conditions: A symposium held at the Pymatuning Laboratory of Ecology 1976. Univ Pittsburgh: 192–205
Waterhouse JA (1974) Cancer handbook of epidemiology and prognosis. Churchill Livingstone, Edinburgh
Watt K (1971) Ecology and management of natural resources. Mir, Moscow
Wawell W (1869) The history of inductive sciences. St Petersburg 3
Williamson M (1975) The analysis of biological populations. Mir, Moscow
Whittaker ET, Watson GN (1927) A course of modern analysis. Univ Press, Cambridge
Wittacker R (1975) Communities and ecosystems, 2nd edn. Macmillan, New York
Woodger JH (1937) Über die Organisatoren in der tierischen Entwicklung
Yablokov AV, Baranov AS, Valetsky AV et al. (1976) The population structure. In: The sand lizard. Nauka, Moscow: 273–287
Yablokov AV, Baranov AS, Rozanov AS (1980) Population structure, geographical variation and microphylogenesis of the sand lizard (*Lacerta agilis* L.). Evol Biol 12:91–127
Yablonsky AI (1977) Scholastic models of research work. In: System investigations. Nauka, Moscow: 5–52
Yakovlev NN (1922) Extinction of animals and its causes according to geological data. Izv Geol Kom 41(1):17–31
Yushkevich AP (ed) (1970) History of mathematics. I. Nauka, Moscow
Zaguskin SL (1984) Evolutionary hierarchy of sensory systems biorhythms. In: Sensory of fish physiology. Kolsky Filial of the USSR Acad Sci, Apatity: 9–16
Zakharov AA (1975) The termite *Anacanthotermes ahngerianus* as a component of entomocomplex of saxaul forest. In: Insects as components of biogeocenosis of saxaul forest. Nauka, Leningrad: 186–206
Zakharov AA (1978) Ant, family, colony. Nauka, Moscow
Zakhvatkin AA (1949) Comparative embryology of lower invertebrates (sources and ways of forming individual development of multicellular organisms). Sov Nauka, Moscow
Zavadsky KM (1961) The theory of species. Izdatelstvo LGU, Leningrad
Zavadsky KM (1966) Principal forms of organization of living matter and their subdivisions. In: Philosophical problems of modern biology. Nauka, Leningrad, Moscow: 29–47
Zavadsky KM (1968) Species and speciation. Nauka, Leningrad
Zavadsky KM, Kolchinsky EI (1977) Evolution of evolution: historicocritical articles on the problem. Nauka, Moscow, Leningrad
Zenkevich LA (1951) Fauna and biological productivity of the sea. Sov Nauka, Moscow

Zenkevich LA (1967) Materials to comparative biogeocenology of land and ocean. Zh Obshch Biol 28(5):523–537
Zenkevich LA, Filatova ZA, Zatsepin VI (1948) LA Zenkevich, Biology of the northern and southern seas of the USSR. Selected works. I. Nauka, Moscow: 121–126
Zharkova VK (1973) Specificity of wood and partly-wood steppe populations of the sand lizard in Ryazanskaya region. In: Problems of herpetology. Author's abstract of report on the 3rd Herpetol Conf. Nauka, Leningrad: 81–82
Zharov VR (1975) Ecology of the marmot *Marmota camtchatica* Doppelmayer from the Barguzin ridge. Author's abstract of candidate's dissertation. Irkut Univ, Irkutsk
Zhirmunsky AV (1966) Problems of cytology. In: Handbook of cytology, vol 2. Nauka, Moscow, Leningrad: 623–637
Zhirmunsky AV, Kuzmin VI (1980) Critical periods in ontogenesis of man and maximal duration of life. Dokl Acad Nauk SSSR 254(1):251–253
Zhirmunsky AV, Kuzmin VI (1982) Critical levels of development of biological systems. Nauka, Moscow
Zhirmunsky AV, Kuzmin VI (1983) Critical levels in species structure of bottom communities as illustrated by invertebrates of the Vostok Bay (Sea of Japan). Sov J Mar Biol (Engl Transl Biol Morya) 9(2):57–69
Zhirmunsky AV, Kuzmin VI (1986) Critical levels of development of natural systems. Preprint of the Inst Biol Morya, Dalnevost Nauch Center Acad Nauk SSSR, Vladivostok
Zhirmunsky AV, Kuzmin VI, Nalivkin VD, Sokolov BS (1980) Modelling of critical boundaries in development of systems and "periodization" of the Earth history. Preprint of the Inst Biol Morya, Dalnevost Nauch Center Acad Nauk SSSR, Vladivostok
Zhirmunsky AV, Kuzmin VI, Yablokov AV (1981) Critical levels of development of population systems. Zh Obshch Biol 42(1):19–37
Zhirmunsky AV, Oshurkov VV (1984) On species structure of marine foulings and its change in a process of succession. In: Macroevolution. Mat I Vses Konf po problemam evolutsii. Nauka, Moscow
Zhirmunsky AV, Zadorozhny IK, Naidin DP (1967) Determination of growth temperatures of some recent and fossil molluscs by the ratio O^{18}/O^{16} in their skeletal formations. Geokhimiya 5:543–552
Zimina AM (1956) On physiological principles of compensation of disturbed functions. Medgiz, Moscow
Zipf GK (1949) Human behaviour and the principle of least effort. Univ Press, Cambridge
Zolotaryov VN (1974) Many-year growth rates of mussels *Crenomytilus grayanus* (Dunker). Ekologiya 3:76–80
Zotin AI (1979) The function of connected dissipation – the divergence of data of direct and indirect calorimetry in living organisms. Zh Obshch Biol 40(2):229–273

Subject Index